ENVIRONMENTAL
AND HYDRAULIC
ENGINEERING

LABORATORY MANUAL

Gang Chen
Simeng Li
Youneng Tang

J.ROSS
PUBLISHING

Contents

Preface

This laboratory manual is composed of 14 laboratory experiments, covering topics of water quality, water treatment, groundwater hydrology, liquid static force, pipe flow, and open channel flow. These experiments are organized with a deliberately logical flow to cover the related topics.

This manual is divided into two sections: environmental engineering experiments and hydraulic engineering experiments, with seven experiments for each section. It provides the basic hands-on training for junior level civil and environmental engineering students. In each experiment, fundamental theories in the topic area are revisited and mathematic equations are presented to guide practical applications of these theories. Tables, figures, graphs, and schematic illustrations are incorporated to give a better understanding of concept development, experimental design, and data collection and recording. Each experiment ends with discussion topics and questions to help students further think and understand the content of the experiment.

This manual mainly serves as a textbook for a laboratory course of environmental engineering and hydraulics engineering. Professionals and water/wastewater treatment plant managers may also find this manual handy for their daily jobs. In addition, students in related areas can use this manual as a reference and the general public may use it to educate themselves on water quality testing and water flow.

About the Authors

Dr. Gang Chen currently is a professor and the program coordinator for water resources and environmental engineering at Florida A&M University-Florida State University (FAMU-FSU) College of Engineering. He finished his Ph.D. at the University of Oklahoma in 2002. Before that, he studied at Harbin Institute of Technology and Cornell University. He joined Florida State University in 2005 after working three years as a postdoctoral research associate at Washington State University. Dr. Chen has been teaching in the area of water resources and environmental engineering for over 10 years. He integrated the Water Resources Engineering Laboratory and the Environmental Engineering Laboratory into a single course and developed this essential manual. He has published over 70 technical journal papers.

Mr. Simeng Li is a Ph.D. student in the Department of Civil and Environmental Engineering of FAMU-FSU College of Engineering. He joined the department in 2016 and started teaching the Water Resources and Environmental Engineering Laboratory ever since.

Dr. Youneng Tang is an assistant professor at Florida State University. He obtained his Ph.D. from Arizona State University in 2012 and joined the Department of Civil and Environmental Engineering of FAMU-FSU College of Engineering in 2015 after working as a postdoctoral research associate at University of Illinois at Urbana-Champaign from 2012–2015.

Laboratory Safety

Safety is always a fundamental topic for all laboratory activities. All levels of administrative and academic management are responsible for promoting safety and well-being in the laboratory. Instructors who have direct contact with students in the teaching environment must ensure the safety of students and keep them away from hazards. In the meantime, students must obey laboratory safety requirements to ensure safety for themselves and their classmates.

While conducting experiments in the laboratory, there is a danger of injury or even death resulting from accidental contact with hazardous material, uncontrolled reactions, explosions, fire, and/or electrical shock. Therefore, it is important for both the instructors and students to know how to minimize risks and what to do in case of an accident.

GENERAL GUIDELINES

- Students are not allowed to work in any laboratory without the presence of the instructor.

- Students must exactly follow directions and laboratory procedures, unless stated otherwise by the instructor.

- Students may perform only the experiments authorized by the instructor, carefully following all guidance and instructions. Unauthorized experiments are not allowed.

- Students should clean and dry the work area before leaving the laboratory and discard scrap materials in designated places.

- The instructor must make the locations of eyewash stations, first-aid kits, fire extinguishers, fire alarms, and proper exit routes known to the students.

- Eating, drinking, and smoking are strictly prohibited in the laboratory.

- All chemicals must be disposed of into appropriate, designated receptacles per the directions given by the instructor.

- Anyone violating any rules or regulations may be denied access to these facilities.

SAFETY PRECAUTIONS

- Medical conditions of any kind must be reported prior to the laboratory work.

- Close-toed shoes must be worn in the laboratory.

- Hair and loose clothing should be fastened or secured.

- Do not wear loose jewelry that can be entangled in power tools or cause electrification.

- Do not wear neckties while working around machinery.

- Never use a tool that you are not familiar with.

- Disconnect the power source prior to performing maintenance and/or cleaning.

- Read Material Safety Data Sheets (MSDS) for safety information before using the material, and use personal protective equipment as necessary.

- Do not use the heat exhaust hood without authorization.

EMERGENCY RESPONSE PROCEDURES

- Be prepared to act intelligently and quickly. Immediate action must be taken to prevent injury or property damage.

- Notify the instructor or someone nearby immediately in the event of any accident, no matter how minor it may be.

- If corrosive liquids (acidic or basic) touch the skin, flood the area with water first and then notify the instructor. This rinsing should last at least 5 minutes.

- Chemicals in the eye(s) should be rinsed with the eyewash immediately for 15 minutes. Notify the instructor after finishing.

Instructors may have additional rules for specific situations or settings. All safety related information and procedures must be stated ahead of time. Students must abide by all verbal and written instructions in carrying out the activity or investigation.

I have read the above section and agree to comply with all the safely requirements while participating in each laboratory experiment.

_____

Student name (printed)

_____   _____

Student name (signature)   Date

Laboratory Reports

One of the essential aspects of engineering jobs is to clearly and effectively communicate ideas, developments, and results to other engineers and their managers through writing and documenting their work. Preparing technical documentation often occupies more than half of the engineers' working time. Therefore, training in documentation and technical communication is essential in engineering curricula. This laboratory course provides the students with opportunities to practice technical writing through laboratory report composition.

Though the topics and specific requirements may vary, the following items or subsections are expected to be included in the laboratory report:

1. Title page

2. Table of contents

3. Abstract

4. Introduction (including background and theoretical discussion)

5. Experimental procedures and methodologies

6. Experimental results and discussion

7. Conclusion or summary

8. Acknowledgements

9. Appendices

10. References

When turning in the laboratory reports, data sheets with recorded raw data need to be torn out from this manual and turned in as appendices to the reports.

PART 1

Environmental Engineering Experiments

Experiment 1
Turbidity Measurement

OBJECTIVE

The objective of this experiment is to generate a calibration curve that will be used to determine the turbidity of a water sample. This experiment also outlines the usage of linear regression techniques in determining the calibration curve and is designed for the students to get familiar with turbidity measurement and linear regression applications.

INTRODUCTION

Turbidity is a measure of the cloudiness or haziness of water. In natural waters, turbidity is often caused by suspended solids (mainly soil particles) and plankton (microscopic plants and animals) that are suspended in the water (Russell, 1994). Moderately low levels of turbidity with appropriate amount of plankton may indicate a well-functioning healthy ecosystem, but high levels of turbidity pose environmental problems (Berger and Argaman, 1983). The colloidal and finely dispersed turbidity-causing materials absorb and scatter light. Therefore, dissolved oxygen may be low in turbid waters when the light needed by the submerged aquatic vegetation for photosynthesis is blocked out. Turbidity can also raise the temperature to abnormal levels in surface waters, as more heat from the sunlight is absorbed by the suspended particles. In addition, suspended soil particles may carry pollutants such as excessive nutrients and pesticides throughout the stream systems, resulting in plankton blooms and threats to the environmental and public health (Gilvear, 1987; Boyd and Tucker, 1998).

Turbidity in water is undesirable because it can cause water to be aesthetically unpleasant. It can also serve as indicative evidence of the presence of bacteria in water. Currently, turbidity is the most important parameter in water supply engineering and its measurement is commonly performed using a proprietary nephelometric instrument and is expressed as Nephelometric Turbidity Units (NTU) (Russell, 1994). According to the Maximum Contaminant Level (MCL) of Drinking Water Standards of the U.S. Environmental Protection Agency (EPA), public drinking water from systems that use flocculation or direct filtration for turbidity control cannot exceed 1.0 NTU when leaving the treatment plant (Eaton *et al.*, 2005). In the samples collected for turbidity measurement, the turbidity should remain no higher than 0.3 NTU for at least 95% of those collected in any month. If a public drinking water system uses any filtration other than flocculation or direct filtration, the turbidity is then subject to individual state limit. However, the turbidity must not exceed 5.0 NTU (Russell, 1994).

In practice, turbidity is measured by a turbidity meter. During measurement, a light beam passes through the water sample, and the light energy is absorbed and re-radiated by the particles in the water sample. The light transmitted through the sample in the forward direction heavily depends upon the particle size in the water sample. To eliminate the influence of particle size, turbidity is usually measured at a certain angle to the incident light beam (Andreadis, 2002). The International Standard (ISO7027:1999(E)) and North American EPA Method (1801.1) require the measurement of turbidity at 90° to the incident light beam, which can be accomplished by a turbidity meter composed of a tungsten-filament lamp, a 90° detector to monitor the scattered light, and a transmitted light detector (Eaton *et al.*, 2005) (Figure 1.1). The commonly used turbidity meters or Nephelometers rely on a photoelectric detector to measure the forward light scattering at 90° to the path of an incandescent light beam and then compare the readings against the light reflected by a reference standard (Eaton *et al.*, 2005).

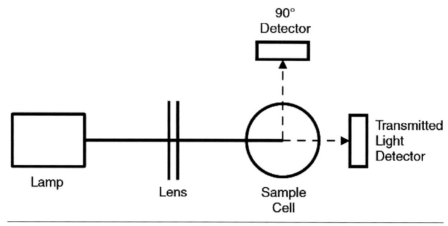

Figure 1.1   Schematic of turbidity measurement in a turbidity meter

Water turbidity in NTUs has a linear relationship with light scattering, or more straightforwardly, the turbidity meter readings. Therefore, a linear regression line needs to be generated to transfer the turbidity meter readings (light scattering) to water turbidity.

From linear regression, water turbidity can be predicted based on turbidity meter readings. The turbidity that will be predicted is called the criterion variable and is referred to as Y. The turbidity meter reading is called the predictor variable and is referred to as X. Since there is only one predictor variable, the prediction method is simple linear regression. In simple linear regression, the prediction of Y, when plotted as a function of X, forms a straight line.

Using a set of example data in Table 1.1, a positive linear relationship between X and Y is plotted in Figure 1.2. On the chart, right click the data series and select "Add Trendline…" (Figure 1.3). Then select "Linear" and enable "Set Intercept", "Display Equation on the Chart" and "Display R-Squared Value on Chart" (Figure 1.4).

The obtained linear regression consists of the best-fit straight line through the points, which is called a regression line. The dotted diagonal line in Figure 1.4 is the regression line and consists of the predicted score on Y for each possible value of X. The specific relationship between Y and X is described by the equation circled in Figure 1.4. Using this equation, Y values can be predicted based on X values.

MATERIALS

1. Hach 2100N Turbidity Meter

    The 2100N Turbidity Meter is equipped with a stable halogen-filled tungsten filament lamp to meet the reporting requirements of EPA Method 180.1 (Figure 1.5). This model is for basic laboratory testing of water samples up to 4,000 NTU.

Table 1.1   Example data of simple linear regression

X	Y
0	0
1.00	1.00
2.00	2.00
3.30	3.00
3.75	4.00
5.25	5.00

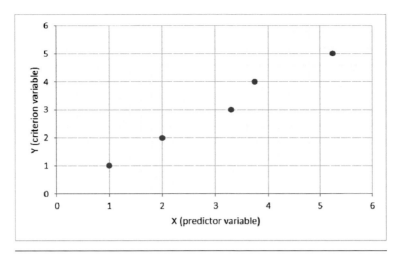

Figure 1.2   Scatter plot of the example data

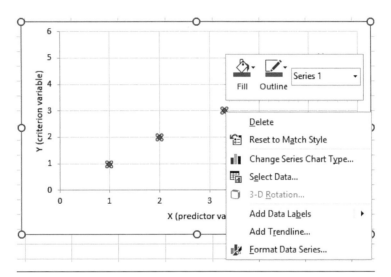

Figure 1.3   Adding a trendline to the plotted data

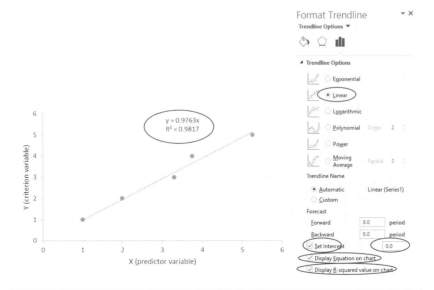

Figure 1.4   Linear regression and display of R-squared

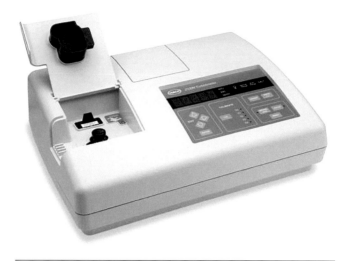

Figure 1.5   Hach 2100N Turbidity Meter

Figure 1.6   Turbidity standards kit for Hach 2100N Turbidity Meter

2. Turbidity standards kit for Hach 2100N Turbidity Meter

    • Standard-filled cuvettes with the NTU values labeled on the caps (Figure 1.6)

    • Clean empty glass cuvettes

3. Water sample with unknown turbidity

PROCEDURES

1. Fill the clean cuvette with deionized water to the line (≅30 mL). Wipe the outside of the cuvette with Kimwipes to remove water spots and finger prints.

2. Place the cuvette into the cell holder, and close the cell cover.

3. Press the "ZERO" key to calibrate the turbidity meter.

4. Next, place the first standard cuvette into the cell holder and press the "ENTER" key. Record the turbidity reading in Table 1.2. Then take the standard cuvette out, rotate the cuvette 90 degree and place it into the cell holder and re-do the reading. Do this again for rotations of 180 and 270 degrees. Quadruple readings will be thus obtained for the standard.

5. Repeat the above step for all the standard cuvettes and record the corresponding turbidity readings in Table 1.2.

6. Generate the linear regression of turbidity by plotting labelled values on each standard cuvette on the y-axis versus the average of the four turbidity readings on the x-axis following the steps described before.

7. Fill the clean cuvette with the water sample with unknown turbidity to the line ($\cong$30 mL). Wipe the cuvette with Kimwipes to remove water spots and finger prints. Place it in the cell holder and close the cell cover. Press the "ENTER" key. Record the turbidity meter reading in Table 1.2.

8. Fit the turbidity meter reading from Step 7 into the linear regression equation generated from Step 6 to obtain the turbidity of the water sample.

NOTE: the derivation of the linear regression for a calibrated Hach 2100N Turbidity Meter is not typically required. A calibrated Hach 2100 N Turbidity Meter can be used to measure the turbidity of a sample by only performing Step 7. This lab is designed for you to understand the calibration process.

DISCUSSION

1. During linear regression, you see an option of "Set Intercept", which is set to "0.0". Compare the results (turbidity of the samples) you obtain between enabling and disabling "Set Intercept". Do you see any difference? In which situation should you disable this option?

2. What is the meaning of R^2 associated to the linear regression line? How does this value affect your results?

DATA RECORDING*

Table 1.2 Turbidity meter readings of standards and samples

Standards		Turbidity Meter Reading
1	NTU	
1	NTU	
1	NTU	
1	NTU	
2	NTU	
2	NTU	
2	NTU	
2	NTU	
3	NTU	
3	NTU	
3	NTU	
3	NTU	
4	NTU	
4	NTU	
4	NTU	
4	NTU	

Water Sample	Turbidity Meter Reading
Test 1	
Test 2	
Test 3	
Test 4	

*Quadruple readings are required for all the measurements in this experiment. Statistical analysis is required when reporting the final results in the laboratory report.

Experiment 2

Color Measurement

OBJECTIVE

This experiment introduces another major parameter of drinking water quality—water color—and its measurement using the spectrophotometric method. The objective of this experiment is to perform turbidity and color tests on a given set of water samples and examine their progressive change as the water flows from one treatment unit to another at a water treatment facility. Linear regression is used to determine the color and turbidity of the water samples.

INTRODUCTION

Surface waters are colored, primarily due to the presence of organic decomposition products (e.g., humic and fluvic acids), impurities of minerals (e.g., iron and manganese oxides), or colored clays (Hendricks, 2005). This color is considered *apparent color* as it is observed in the presence of suspended matters (turbidity) that naturally occur and exist, while *true color* results only from dissolved organic and inorganic matters (Eaton et al., 2005). In order to measure true color, samples are centrifuged and/or filtered to remove the turbidity.

Overall color level may be an indicator of the organic content in the water. Waters that contain color from natural organic matters usually pose no health hazard. However, because of the yellowish or brown appearance, consumers may find them aesthetically unacceptable. Both color and turbidity removal are therefore required for domestic water usage. National secondary drinking water regulations (NSDWRs) by the U.S. Environmental Protection Agency (EPA) recommend a non-enforceable color level of 15 color units (USEPA, 1977).

Color in the water limits the penetration of light. Subsequently, color levels of the water samples can be measured by a spectrophotometer, given appropriate color standards. EPA Method 110.2 (i.e., the Platinum-Cobalt method) is the approved standard method for color measurement (Eaton et al., 2005). In this method, one unit of color is equivalent to that produced by 1 mg/L platinum standard solution in the form of the chloroplatinate ion.

When a light beam passes through the water sample, the reflected color intensity is proportional to the color of the water sample measured by a spectrophotometer. By comparison of the reflected color intensity with a color standard, the spectrophotometer reading can be converted to the color unit. For colorimetry measurement, the UV and visible regions of the spectrum can be employed. Common spectrophotometers use the visible region of 400–700 nm. To minimize optical interferences, color measurement is usually conducted with the sample contained in a cuvette that is fabricated by quartz. Glass or plastic is used if less accuracy is acceptable.

MATERIALS

1. Hach DR6000 benchtop spectrophotometer

    The UV-Vis spectrophotometer DR6000 (Figure 2.1) delivers top performance for both routine laboratory tasks and demanding photometry applications. It offers high-speed wavelength scanning across the UV and visible spectrum, and comes with over 250 pre-programmed methods, including color measurement. Note that clean spectrophotometer cuvettes are needed for holding the water samples.

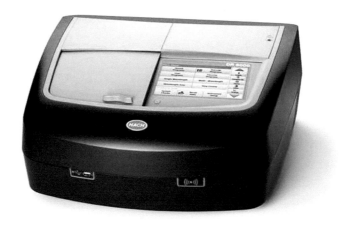

Figure 2.1   Hach DR6000 Benchtop Spectrophotometer

2. Nessler tubes

    Matched, tall-form, 50-mL capacity with stoppers (13 for standards and 3 for water samples)

3. Reagent

    Potassium chloroplatinate (K_2PtCl_6) will be used as the reagent to make the color standards. To make the chloroplatinate solution, dissolve 1.246 g potassium chloroplatinate (equivalent to 0.500 g metallic Pt) and 1 g crystalline cobaltous chloride ($CoCl_2 \cdot H_2O$) in 1000 mL deionized water containing 100 mL of concentrated HCl (38%). Cobaltous chloride helps to match the color between the standards and most natural waters, and the acid helps to dissolve the crystals. This standard solution is equivalent to 500 color units.

4. Standards

    Prepare the color standards from 0 to 70 units in an increment of 5 using standard chloroplatinate solution created above in the Nessler tubes at ratios suggested in Table 2.1. These color standards will be used as references for the determination of the apparent color for the water samples. The standards must be protected against evaporation with clean, inert stoppers. Also, the absorption of ammonia must be avoided since it may lead to an increase in color.

Table 2.1   Suggested color standard dilution series

Chloroplatinate Solution (mL)	Deionized Water (mL)	Color Standard (Chloroplatinate Unit)
0.0	50.0	0
0.5	49.5	5
1.0	49.0	10
1.5	48.5	15
2.0	48.0	20
2.5	47.5	25
3.0	47.0	30
3.5	46.5	35
4.0	46.0	40
4.5	45.5	45
5.0	45.0	50
6.0	44.0	60
7.0	43.0	70

5. Water samples

Typically, raw water (e.g., surface water) will go through a series of treatment processes before it is available as tap water for end users (see Figure 2.2). The lab instructor will collect water samples at the end of each corresponding process unit and mark them as *raw water, settled water, filtered water,* and *tap water,* respectively (shaded in Figure 2.2). Students will measure the color and turbidity of these water samples in this experiment.

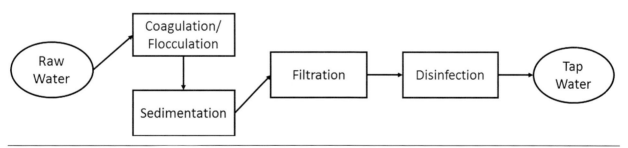

Figure 2.2   Typical flow diagram of a water treatment plant

PROCEDURES

1. Turbidity measurement

Follow the procedures of Experiment 1 for the measurement of the samples' turbidity. **Note that the turbidity can be read directly from the calibrated turbidity meter without the generation of a calibration curve.** Record the results in Table 2.3.

2.   Apparent color measurement

The color of the samples is measured by filling a Nessler tube to the 50-mL mark and comparing the samples with the standards. The comparison is made by looking vertically downward through the tubes toward a white or specular surface placed at such an angle that light is reflected upward through the column of the liquid. The color is reported as *apparent color* before the turbidity is removed by centrifuging the samples. If the color exceeds 70 color units, dilute the samples with deionized water until the color is within the range of the standards. **Note that if dilution is performed, the color should be the reading multiplied by the dilution factor.** Record the results in Table 2.3.

3.   True color measurement

True color will be measured directly using the Hach DR6000 after removing turbidity by centrifuging the samples. The time and speed required for centrifugation depends upon the nature of the samples, mainly the density and particle size. The centrifuge tubes containing the samples are placed in a Sorvall RC-5C Centrifuge (Block Scientific, Inc.) and centrifuged at 16°C. The separation is operated following Stokes' Law, which provides the theoretical basis for the determination of the centrifugation speed. For this experiment, the method developed by M. L. Jackson is used (Jackson, 1969):

$$T_m = \frac{63.0 \cdot 10^8 \eta \, log_{10}\left[\dfrac{R}{S}\right]}{N^2 D^2 \Delta s}$$

where T_m is the centrifugation time, η is the viscosity of water in poise at the existing temperature (0.01111 poise for water at 16°C), R is the radius from the axis of rotation to the top of the sediment in the centrifuge tube or bottle (Figure 2.3), S is the radius from the axis of rotation to the surface of the suspension in the centrifuge tube or bottle, N is the revolutions per minute, D is the particle diameter, and Δs is the specific gravity of the particle. Based on this equation, the centrifugation time required for different-sized particles are listed in Table 2.2. Please note that Figure 2.3 is for the illustration of the dimensions of the centrifugation tubes. The centrifugation tubes will be put vertically at certain angles in the centrifuge, instead of horizontally as shown in Figure 2.3.

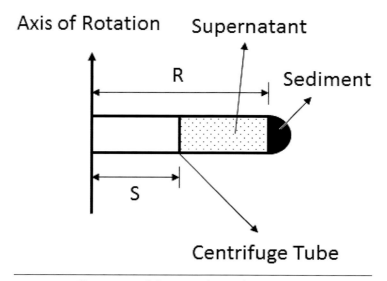

Figure 2.3   Illustration of the centrifuge tube

Table 2.2 Centrifugation time for variable particle sizes

Particle Size	Sedimentation Time at 100 rpm (min)
600 μm (Sieve No. 30)	1.5
300 μm (Sieve No. 50)	3.0
150 μm (Sieve No. 100)	6.0
75 μm (Sieve No. 200)	12.0

The particles that can interfere with water sample true color measurements are usually larger than 75 μm. Therefore, centrifugation of the water samples at 100 rpm and 12 minutes should remove them. After centrifugation, the supernatants will be transferred to the cuvette of the Hach DR6000 and the true color will be read directly from a properly calibrated Hach DR6000 Benchtop Spectrophotometer. If color exceeds 70 color units, dilute the samples with deionized water until the color is within the range of the standards. Note that if dilution is performed, the true color should be the reading multiplied by the dilution factor. Record the results in Table 2.3.

DISCUSSION

1. Discuss the progressive change in the color and turbidity of the samples as the water flows from one treatment unit to another at a water treatment facility (Figure 2.2).

2. Discuss graphically the relationships between turbidity and apparent color and between turbidity and true color.

3. Discuss any possible errors or chance of interference in this experiment.

DATA RECORDING*

Table 2.3 Turbidity and color of water samples from different process units

Sample	Turbidity (NTU)			Apparent Color (Color Unit)			True Color (Color Unit)		
	1	2	3	1	2	3	1	2	3
Raw water									
After sedimentation									
After filtration									
Tap water									

*Triple readings are required for all measurements in this experiment. Statistical analysis is required when reporting the final results in the laboratory report.

Experiment 3

Measurement of pH and Alkalinity

OBJECTIVE

In this experiment, pH and alkalinity analyses will be conducted on a set of water samples from different water treatment processes.

INTRODUCTION

pH is the numeric quantification of the acidity or basicity of an aqueous solution. It is defined as the negative logarithm of the activity of hydrogen ion. In dilute solutions, the activity of hydrogen ion is approximately its molar concentration (Morel and Hering, 1993):

$$pH = -log_{10}\left[H^+\right]$$

In general, an aqueous solution with a pH < 7 is considered acidic, while a pH > 7 is considered basic. The normal pH range is 6.5 to 8.5 for surface water and 6 to 8.5 for groundwater.

Alkalinity is the capacity of an aqueous solution to resist the change in pH that would otherwise make the water more acidic. By buffering water against a change in pH, alkalinity supports microbial functions in the aqueous system. In typical surface water or groundwater, the total alkalinity, A_T is equal to (Morel and Hering, 1993):

$$A_T = \left[HCO_3^-\right] + 2\left[CO_3^{2-}\right] + \left[OH^-\right] - \left[H^+\right]$$

In most natural waters, $[HCO_3^-]$ and $[CO_3^{2-}]$ are much higher than $[OH^-]$ and $[H^+]$. Thus the total alkalinity is approximately equal to $[HCO_3^-] + 2[CO_3^{2-}]$.

The pH and alkalinity of a solution are related, but represent different properties. Roughly, the pH of a solution measures the acidity or basicity of a solution, whereas the alkalinity measures the buffering capacity of the solution to neutralize strong acids.

The pH of a water sample can be determined using a pH meter (Figure 3.1), which measures the electrical potential of the solution by a pH electrode and provides the reading after comparing with that of a reference electrode. The electrical potential is then used to calculate the hydrogen ion concentration, which is further used to calculate the pH. To obtain correct and precise results, pH meters require frequent calibration.

Alkalinity of a water sample is determined by titration (also known as titrimetry), a common laboratory method of quantitative chemical analysis. Sulfuric acid or hydrochloric acid with a known concentration is added to the sample during titration, which continues until the pH of the sample reaches an end point. The volume of the acid added is proportional to the alkalinity of the sample. Two pH end points are used during titration and correspond to two types of alkalinity respectively: pH 8.3 for phenolphthalein alkalinity and pH 4.5 for total alkalinity. When the pH is above 8.3, the following two reactions occur (Morel and Hering, 1993):

$$OH^- + H^+ \rightarrow H_2O$$

$$CO_3^{2-} + H^+ \rightarrow HCO_3^-$$

When the pH is between 4.5 and 8.3, the following reaction occurs (Morel and Hering, 1993):

$$HCO_3^- + H^+ \rightarrow H_2CO_3$$

pH indicators are used to determine whether the pH reaches the end point (8.3 or 4.5) by observing the change of the color of the sample.

Figure 3.1 PICO pH Meter (LABINDIA)

MATERIALS

1. pH meter: PICO pH meter or other models

2. Electrolyte solution for pH electrodes

3. Buffer solutions (4.0 and 7.0)

4. Phenolphthalein indicator solution and bromcresol green indicator solution

5. Sulfuric acid (H_2SO_4): 0.02 N

6. Titration apparatus: burette, burette clamp, rubber stopper, Erlenmeyer flask (100 mL)

7. Five sets of water samples, each containing *raw water*, *settled water*, *filtered water*, and *tap water* from water treatment processes (the same as in Experiment 2)

8. Deionized water

9. Stir plate

PROCEDURES

1. Calibration of pH meter

    The electrode of the pH meter needs to be calibrated regularly. Before calibration, make sure the pH electrode has been saturated in the electrolyte. Turn on the device by pressing the "I/O" key. Make sure the pH meter is in the pH mode. If not, press the "MODE" key until the pH mode is displayed. Press the "CAL" button to enter calibration mode. The display panel will flash. Rinse the electrode with deionized water and place it into the 4.0 buffer. Press "CAL" button. "CALIBRATE" and a calibration slope will be displayed, and "P1" (i.e., point 1) will show in the lower field, which indicates that the meter is ready to accept the first buffer point.

    When the electrode is stable, the meter beeps and "READY" is displayed along with the temperature corrected value for the buffer (flashing). Press the "YES" key to accept this point.

    The display will remain fixed momentarily. Then "P2" is displayed in the lower field, suggesting that the meter is ready to accept the second buffer point (7.0).

    Rinse the electrode with deionized water and place it in the second buffer (7.0). Wait until the meter beeps and "READY" is displayed. Press the "YES" key to accept the second point.

    The pH meter will now automatically enter the measure mode and "MEASURE" will appear in the main display. Now the meter is ready for pH measurement. The meter retains this calibration until a new calibration is entered or until power is disconnected from the instrument.

    Before each measurement, rinse the electrode tip with deionized water. Wait until the reading becomes steady with "READY" displayed and the pH meter beeping once. Record the results for each water sample.

2. Measuring alkalinity

    Add 100 mL of a sample into a 250-mL Erlenmyer flask. Place the sample onto a stir plate and make sure a magnet bar is placed in the flask.

    Measure the initial pH of the sample. If the sample pH is above 8.3, add three drops of phenolphthalein indicator and the color of the sample will turn pink. Titrate the sample with 0.02 N H_2SO_4 until the color changes from pink to clear. Record the volume of acid used for the titration in Table 3.1. Then, add three drops of bromcresol green indicator and the color of the sample will turn blue. Titrate the sample with 0.02 N H_2SO_4 until the color changes from blue to yellow. Record the volume of acid used for the titration in Table 3.1.

    If the sample pH is below 8.3, perform the titration with the bromcresol green indicator directly.

    Measure the alkalinity for the remaining samples and record results in Table 3.1.

3. Calculate phenolphthalein alkalinity and total alkalinity for all samples using the following formulas and record results:

$$Phenolphthalein\ alkalinity \left(as \frac{mg}{L} CaCO_3 \right) = \frac{A \times N \times 1000 \times 50}{mL\ of\ sample} = \frac{A \times 0.02 \times 1000 \times 50}{100}$$

where A is the volume (mL) of 0.02 N H_2SO_4 used for phenolphthalein end point titration and N is the normal concentration of H_2SO_4 used for the titration, which is 0.02 N.

$$Total\ alkalinity \left(as \frac{mg}{L} CaCO_3 \right) = \frac{B \times N \times 1000 \times 50}{mL\ of\ sample} = \frac{B \times 0.02 \times 1000 \times 50}{100}$$

where B is the total volume (mL) of 0.02 N H_2SO_4 used for both phenolphthalein end point titration and bromcresol green end point titration.

DISCUSSION

1. What are the dominant species (CO_3^{2-}, HCO_3^-, OH^- and H^+) in the samples that are used in this experiment?

2. Assuming you observe the following three scenarios, what are the dominant species (CO_3^{2-}, HCO_3^-, OH^- and H^+) in the sample in each scenario and why? a) The pH of a sample is above 8.3. The volume of H_2SO_4 used for both end point titrations are the same and above zero. b) The pH of the sample is above 8.3. H_2SO_4 addition is required to reach the pH end point of 8.3, but not required to further reach the pH end point of 4.5. c) The pH of the sample is between 4.5 and 8.3. A certain amount of H_2SO_4 is required to reach the pH end point of 4.5. Hints: If carbonate (CO_3^{2-}) is present in the sample, 1 mole CO_3^{2-} will consume one mole H^+ when the solution is titrated to pH 8.3, and it will consume another mole H^+ during further titration from pH 8.3 to pH 4.5.

3. Discuss the progressive change in the pH and the alkalinity of the water samples as the water flows from one unit process to another during water treatment.

DATA RECORDING*

Table 3.1 pH and alkalinity of water samples from different unit processes

Sample		Raw Water	After Sedimentation	After Filtration	Tap Water
pH	1				
	2				
	3				
	4				
	5				
Phenolphthalein alkalinity	1				
	2				
	3				
	4				
	5				
Total alkalinity	1				
	2				
	3				
	4				
	5				

*Quintuple readings are required for all measurements in this experiment. Statistical analysis is required when reporting the final results in the laboratory report.

Experiment 4
Determination of Residual Chlorine

OBJECTIVE

The aim of this experiment is to familiarize students with the disinfection process and master the technique to measure the residual chlorine in water after disinfection.

INTRODUCTION

Drinking water is sourced from a variety of places, such as lakes and rivers, with high turbidity, which may contain pathogens that pose a serious risk to human health. During drinking water treatment, turbidity is removed and pathogenic organisms are deactivated by disinfection. The most widely used disinfection method for public water systems is chlorination, which has been practiced for more than 100 years (USEPA, 2009). Although other disinfection technologies like ozonation and ultraviolet (UV) radiation have been well developed, chlorination still remains as the major disinfection means for water treatment.

During chlorination, chlorine is added to water in a gaseous form or as sodium/calcium hypochlorite. Chlorine gas rapidly hydrolyzes to hypochlorous acid, $HOCl$, which then dissociates following the following reactions (Fielding and Farrimond, 1999):

$$Cl_2 + H_2O \rightarrow HOCl + H^+ + Cl^-$$

$$HOCl \rightarrow OCl^- + H^+$$

Hypochlorous acid can react with a wide variety of compounds in water. Of particular importance in disinfection is the reaction of hypochlorous acid with nitrogenous compounds that are commonly present in natural waters, such as ammonia. Hypochlorous acid interacts with ammonia to form monochloramine, dichloramine, and trichloramine, depending on several factors such as pH, temperature, and the chlorine dosage (Fielding and Farrimond, 1999):

$$NH_3 + HOCl \rightarrow NH_2Cl \left(monochloramine\right) + H_2O$$

$$NH_2Cl + HOCl \rightarrow NHCl_2 \left(dichloramine\right) + H_2O$$

$$NHCl_2 + HOCl \rightarrow NCl_3 \left(trichloramine\right) + H_2O$$

The three chloramines are also oxidants and can serve as disinfectants, but they are not as effective as hypochlorous acid or hypochlorite ion, OCl^-.

The break point in chlorination describes the process of the interaction of chlorine with ammonia. At the break point, all demand for chlorine is neutralized and free chlorine ($[Cl_2]$ (dissolved chlorine gas) + $[HOCl]$ (hypochlorous acid) + $[OCl^-]$ (hypochlorite)) begins to accumulate. The difference between total chlorine and free

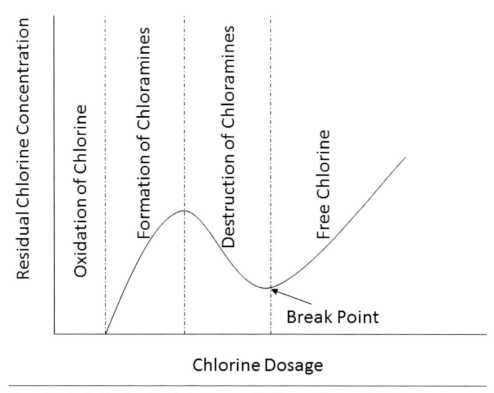

Figure 4.1   Typical break-point chlorination curve (Fielding and Farrimond, 1999)

chlorine is called the combined chlorine, which exists in forms of chloramines. Before the break point is reached, the combined chlorine dominates (Figure 4.1). After the break point, free chlorine becomes the dominant chlorine species in water. Typically, the free chlorine residual is required to maintain a minimum level of 0.2 mg/L for tap water supply.

The chlorine residual in water can be determined using the N, N-diethyl-para-phenylenediamine (DPD) colorimetric method, which was introduced by Dr. Thomas Palin in 1957 (McGuire, 2013). Over the years, it has become the most widely used method for the determination of free and total chlorine in water and wastewater.

Chlorine oxidizes DPD into two oxidation products: Würster dye and an imine compound. At a near neutral pH, the Würster dye, a semi-quinoid cationic compound, is the primary oxidation product, which is relatively stable and accounts for the magenta color in the DPD colorimetric test. Further DPD oxidation forms a relatively unstable, colorless imine compound, resulting in apparent *fading* of the colored solution (McGuire, 2013). For the residual chlorine, only Würster dye will be formed. The Würster dye has a doublet peak with maxima at 512 nm and 553 nm on the absorption spectrum (Figure 4.2).

If chloramines are to be measured, iodide (I^-) is added to react with the chloramines to form triiodide ion (I_3^-), which further reacts with DPD to form the Würster dye. Iodide can also prevent the free chlorine from reacting with the DPD. The amount of iodide added depends on forms of the chloramines. In most water samples, monochloramine is dominant and 0.1 g KI per10 mL sample is added. Phosphate buffer is usually added to maintain the reaction pH.

In this experiment, the wavelength of 515 nm in a Hach DR6000 spectrophotometer will be used for the determination of the Würster dye following the Ultra-Low Range method (Hach, 2016). The Würster dye displays a red color and the intensity of the red color has a linear relationship with the oxidant (chlorine) concentration.

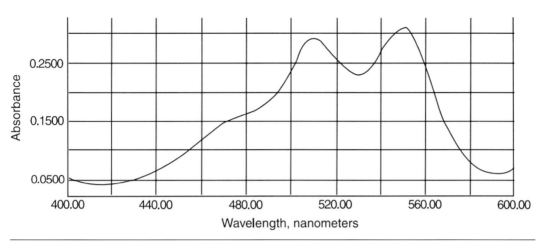

Figure 4.2   Absorption spectrum—DPD Würster compound (Fielding and Farrimond, 1999)

MATERIALS

1. Spectrophotometer: Hach DR6000 or other models with a wavelength of 515 nm

2. Phosphate buffer solution

3. DPD indicator solution

4. Potassium iodide (KI) crystals

5. Standard potassium permanganate solutions

6. Tap water sample

7. Nine 100 mL (type) flasks

8. Six 15 mL test tubes

9. Pipettes

PROCEDURES

Standard Preparation

1. Dilute the potassium permanganate stock solution (89.1 mg/L, equivalent to 100 ppm Cl_2) provided by the instructor to make a series of potassium permanganate standards in 100 mL flaks (Table 4.1). Add 5 mL of phosphate buffer and 5 mL of DPD indicator reagent in each flask and mix thoroughly. These standards will cover the equivalent chlorine range from 0 to 4 mg/L with deionized water serving as the blank (0 mg/L).

Table 4.1   Suggested chlorine standard dilution series

Stock Solution (mL)	Deionized Water (mL)	Chlorine Equivalent (mg/L)
0.0	100.0	0
0.5	95.5	0.5
1.0	99.0	1.0
1.5	98.5	1.5
2.0	98.0	2.0
2.5	97.5	2.5
3.0	97.0	3.0
3.5	96.5	3.5
4.0	96.0	4.0

2. Zero the spectrophotometer at the wavelength of 515 nm using the blank in accordance with the manufacturer's instructions.

3. Generate the standard curve of mg/L equivalent chlorine versus spectrophotometer readings by recording the spectrophotometer readings of each standard.

Residual Chlorine Determination

4. Add 5 mL of phosphate buffer and 5 mL of DPD indicator reagent to 100 mL of tap water in a 100-mL test tube. Mix thoroughly and let stand for two minutes to allow color to develop.

5. Record the spectrophotometer reading of the above water sample in Table 4.2. Based on the standard curve generated in Step 3, residual chlorine can be determined. Repeat Steps 4 and 5 three times to get three separate readings of residual chlorine in the tap water sample and record the results in Table 4.3.

Chloramines Determination

6. Add 0.5 mL phosphate buffer, 0.5 mL DPD indicator reagent, and approximately 0.1 g of potassium iodide (KI) into 100 mL of tap water in a 100 mL test tube. Mix thoroughly and let stand for two minutes to allow color to develop.

7. Record the spectrophotometer reading of the above water sample in Table 4.2. Based on the standard curve generated in Step 3, chloramine chlorine can be determined. Repeat Steps 6 and 7 three times to get three separate readings for the chloramine chlorine in the tap water sample and record the results in Table 4.3.

DISCUSSION

1. Why is residual chlorine required in the tap water? Does residual chlorine in the tap water sample meet EPA and state requirements?

2. What factors may interfere with the residual chlorine determination using the DPD colorimetric method?

DATA RECORDING*

Table 4.2 Standard curve

Concentration of Standard (mg/L)	Spectrophotometer Reading
0.0	
0.5	
1.0	
1.5	
2.0	
2.5	
3.0	
3.5	
4.0	

Table 4.3   Free chlorine residual and chloramine chlorine in tap water samples

	Free Chlorine Residual (mg/L)	Chloramine Chlorine (mg/L)
Tap Water Sample		

*Triple readings are required for all measurements in this experiment. Statistical analysis is required when reporting the final results in the laboratory report.

Experiment 5

Determination of Optimal Coagulant Dosage

OBJECTIVE

In this experiment, jar testing will be conducted to examine coagulation, flocculation, and sedimentation, and to determine the optimal coagulant dosage.

INTRODUCTION

Drinking water treatment removes turbidity contributed by suspended particles from the water. These suspended particles mostly arise from land erosion, dissolution of clay minerals, and decay of plants. Most suspended particles are negatively charged and the repulsion interaction prevents aggregation, resulting in a stable suspension (Droste, 1996). The particle surface charges originate from ionic adsorption, which are surrounded by a cloud of counter-ions, extending from the surface into the solution (Figure 5.1). These ion/counter-ion layers are known

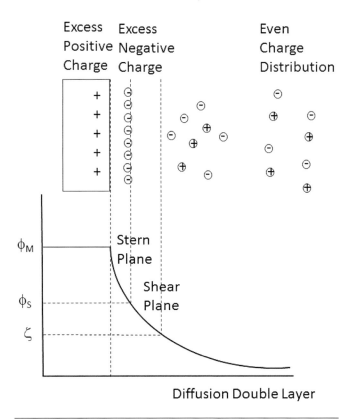

Figure 5.1 A cloud of counter-ions surrounding the particle surfaces. Stern Plane is the plane drawn through the center of the ions adsorbed relatively strongly to the particle surface and Shear Plane is the plane that defines a region at which the fluid becomes mobile and shows elastic behavior (Droste, 1996).

as the electric double layers (Droste, 1996). A solution with a higher concentration of electrolytes decreases the thickness of the double layers.

Coagulation-flocculation is the most commonly used water treatment technique for surface water treatment. Coagulation and flocculation occur in successive steps, which eliminate the repulsive forces that stabilize the suspended particles and allow the particle collision and floc growth. The most widely used coagulant is aluminum sulfate (referred to as alum), which is manufactured from the reaction of bauxite ores with sulfuric acid. The alum used for water treatment has an approximate formula of $Al_2(SO4)_3 \cdot 14-18H_2O$ and Al content ranging from 7.4 to 9.5%.

The most common problem in water treatment is the overfeeding or overdosing of the coagulant. Jar testing is thus practiced to determinate the coagulant dosage by simulating the coagulation-flocculation full-scale operations. Jar testing can also examine alternative coagulants and corresponding dosages without altering the performance of the full-scale treatment process by comparing the formation, size, and settleability of the flocs at the bench scale.

The sedimentation of the flocs is controlled by gravitational, buoyant, and drag forces (Figure 5.2). The vertical forces acting on the flocs are balanced as (Henze *et al.*, 2002):

$$\sum F = F_g - F_b - F_d$$

where F_g is the gravitational force, F_b is buoyant force, and F_d is the drag force.

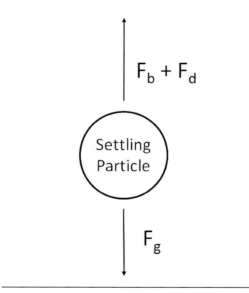

Figure 5.2   Vertical forces that control a settling particle

The gravitational and buoyant forces are given as:

$$F_g = ma = \rho_p V_p g$$

$$F_b = ma = \rho_w V_p g$$

where m is the mass of the particle, a is the velocity acceleration, ρ_p is the density of the particle, ρ_w is the density of water, V_p is the volume of the particle, and g is the acceleration due to gravity.

The drag force F_d can be calculated as (Henze *et al.*, 2002):

$$F_d = C_d \rho_w A_p \frac{v_s^2}{2}$$

where C_d is the drag force coefficient, A_p is the projected area of the particle in the direction of settling, and v_s is the settling velocity of the particle.

Assuming the particle is spherical, the volume and projected area can thus be estimated as:

$$V_p = \frac{\pi}{6} d_p^3$$

$$A_p = \frac{\pi}{4} d_p^2$$

where d_p is the particle diameter.

Thus, the calculated settling velocity is:

$$v_s = \sqrt{\frac{4g\left(\rho_p - \rho_w\right)d_p}{3C_d \rho_w}}$$

In the above equation, the drag force coefficient C_d is a function of the Reynolds number, which is defined as (Henze *et al.*, 2002):

$$Re = \frac{\rho_w v_s d_p}{\mu} = \frac{v_s d_p}{\nu}$$

where Re is the Reynolds number, μ is the dynamic viscosity, and ν is the kinematic viscosity.

For laminar flow ($Re < 2$), $C_d = \dfrac{24}{Re}$. Subsequently,

$$v_s = \frac{g\left(\rho_p - \rho_w\right)d_p^2}{18\mu}$$

MATERIALS

1. Jar test apparatus (Figure 5.3)

2. Six 1500 mL beakers

3. Pipettes

4. Turbidity meter

5. pH meter

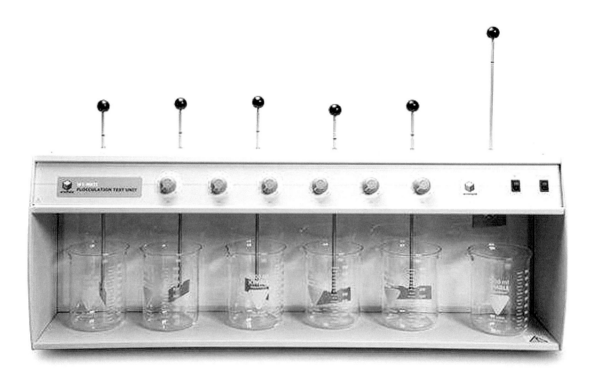

Figure 5.3   Jar test apparatus

6. $Al_2(SO_4)_3 \cdot 14H_2O$ or $FeCl_3 \cdot 6H_2O$

7.   River water samples

PROCEDURES

1.   Prepare a coagulant stock solution by dissolving 10.0 grams of alum or ferric coagulant into 1 L solution with deionized water. Each 1.0 mL of this stock solution contains 10 mg coagulant.

2.   Fill each of the six 1500 mL beakers with 1 L river water.

3.   Measure the initial turbidity and pH of the river water and record the results in Table 5.2.

4.   Add alum or ferric coagulant to each beaker to reach a final concentration of 0, 5, 10, 20, 40, and 100 mg/L, respectively. Refer to Table 5.1 for the stock solution dosage requirements.

Table 5.1   Stock solution dosage requirement

Jar #	mL Alum/Ferric Coagulant Stock Solution Added	mg/L Alum/Ferric Coagulant Dosage
1	0	0
2	0.5	5
3	1.0	10
4	2.0	20
5	4.0	40
6	10.0	100

5. Follow these mixing protocols for each jar:

    a. Rapid mixing (250 rpm) for 1 minute — The rapid mixing helps to disperse the coagulant.

    b. Slow mixing (30 rpm) for 15 minutes — The slow mixing helps promote floc formation by enhancing particle collision. The mixing should be slow enough to prevent the breakup of formed flocs by the mixing turbulence.

    c. Settling for 30 minutes — Settling stage separates the flocs from the solution by gravity.

6. Carefully monitor the floc formation in each beaker during the experiment and write down the observations about the floc formation in Table 5.2.

7. Use a pipette to transfer 30 mL supernatant (from ~ 1 inch below the water surface) to the turbidity meter cuvette to measure the turbidity. Make sure that the pipette does not disturb the water in the beakers. Do this three times for each jar sample, remembering to use a new/clean pipette for each jar. Record results in Table 5.2.

8. Measure the final pH for each jar sample.

DISCUSSION

1. Which beakers have significant floc formation and which beakers do not? Why? Is big floc formation desirable? Why? (Hint: Using mathematical equations to explain)? What is the optimum dosage for this river water?

2. Figures 5.4 and Figure 5.5 display the jar test results of two different water samples treated with two different coagulants, alum and ferric chloride, at varying doses. Water sample I has high alkalinity and requires less coagulant to achieve good coagulation and flocculation. On the contrary, water sample II has low alkalinity and a high dosage of coagulant is required to achieve an acceptable water turbidity. What is the possible explanation for the difference between results from these two samples?

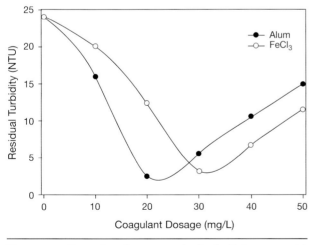

Figure 5.4 Example I of jar test results (high alkalinity sample)

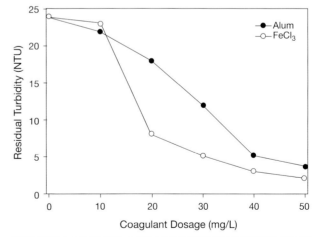

Figure 5.5 Example II of jar test results (low alkalinity sample)

DATA RECORDING*

Table 5.2 Raw data for jar testing

Alum/Ferric Coagulant Dosage (mg/L)	Before Treatment		After Treatment		
	Turbidity	pH	Floc Formation Observation	Turbidity	pH
0					
0					
0					
5					
5					
5					
10					
10					
10					
20					
20					
20					
40					
40					
40					
100					
100					
100					

*Triple readings are required for all measurements in this experiment. Statistical analysis is required when reporting the final results in the laboratory report.

Experiment 6

Sedimentation Characteristics

OBJECTIVE

Solid settling plays a key role for both water and wastewater treatment. In this experiment, the solid settling characteristic curve will be generated for type III settling.

INTRODUCTION

Solid settling refers to the solid-liquid separation by gravitational sedimentation. This is a key step in both water and wastewater treatment. For water treatment, the solids are the coagulated clay minerals; while for wastewater treatment, the solids are the activated sludge. Depending on the size and density of particles, and physical properties of the solids, solid settling in water and wastewater treatment can be classified into four types (Henze *et al.*, 2002).

Type I settling or free settling is the settling of discrete, non-flocculent particles in a dilute suspension. The particles settle as separate units with no apparent interaction between them. Typical type I settling includes the plain sedimentation of surface water and the settling of sand particles in grit removal chambers.

Type II settling is the settling of flocculent particles in a dilute suspension, as observed in the prior experiment. The particles flocculate during the settling, which increase in size and settle at a faster speed. Settling characteristics of typical Type II settling are usually evaluated by batch settling tests.

Type III settling or zone settling is the sedimentation of particles at higher concentrations in which the particles are so close that inter-particle forces hinder the settling of the neighboring particles. The particles remain in a fixed position relative to each other and all settle at a constant velocity. As a result, the mass of particles settles as a zone. There will be a distinct solid-liquid interface between the settling particle mass and the clarified liquid at the top of the settling mass. Typical Type III settling occurs in an intermediate depth in a final clarifier for the activated sludge process.

Type IV settling or compressing settling is the settling of particles that are of such a high concentration that the particles touch each other and settling can occur only by compression of the compacting mass. Typical Type IV settling occurs in a lower depth of a final clarifier for the activated sludge process.

The settling type mainly depends on the solids concentration in the liquid. This experiment simulates the Type III settling in a settleometer (Figure 6.1) and investigates its characteristics by studying the moving velocity of the solid-liquid interface (i.e., interface between the clear liquid and the constant composition zone in Figure 6.1) and the settling flux due to gravity (G_s). The settling velocity is detemined by the traveling distance of the solid-liquid interface over the corresponding traveling time. We would expect that the solids concentration below the interface increases, resulting in changes in the settling velocity. A plot of settling velocity (v) versus solids concentration below the interface (C) will be plotted in the experiment to study how solids concentration affects the settling velocity.

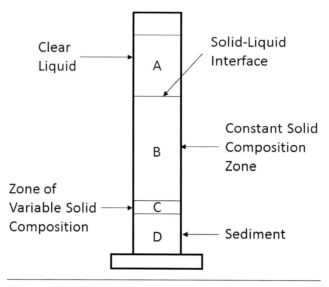

Figure 6.1   Sludge settling in a settleometer

The settling flux due to gravity (G_s) is determined by multiplying the solids concentration (right below the interface, C) by the settling velocity (i.e., traveling velocity of the interface, v):

$$G_s = vC$$

A plot of the G_s versus C will be made in this experiment to study the effect of solids concentration on the settling flux due to gravity.

MATERIALS

1. A settleometer with sampling ports (Figure 6.2)

2. Alum (10.0 g)

3. Timer

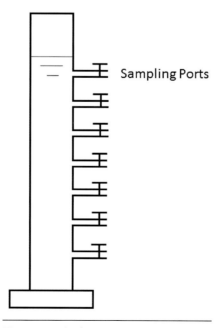

Figure 6.2   Settleometer

PROCEDURES

1. Prepare a solution to simulate water treated by coagulation by dissolving 10.0 grams of alum into a settleometer containing 1 L of deionized water.

2. When a clear sludge-supernatant interface is observed, start to monitor the settling of the interface by recording the time for the interface to move from one sampling port to the next, based on which the settling velocity will be determined and recorded in Table 6.1. Take a 5-mL sample right before the interface arrives at each sampling port by opening the valve of the sampling port. The solids concentration will be calculated by:

$$C = \frac{M_s - M_{DI}}{5\ mL}$$

    where M_S is weight of the 5-mL sample and M_{DI} is the weight of the 5-mL deionized water.

3. Record the data of the above calculations in Table 6.1.

4. Repeat Steps 1-2 two more times.

5. Generate the two settling characteristic curves, i.e., v versus C and G_s versus C using the average of the results.

DISCUSSION

1. Discuss how the solids concentration (C) affects the settling velocity (v) and why?

2. Discuss how the solids concentration (C) affects the solids flux due to gravity (G_s) and why?

DATA RECORDING*

Table 6.1 Sludge-supernatant interface settling

Sampling Port	Distance	Time	v = Distance/Time	C	G_s
1					
2					
3					
4					
5					
6					
7					
8					

*Triple readings are required for the measurements in this experiment. Statistical analysis is required when reporting the final results in the laboratory report.

Experiment 7

Physiochemical Water Treatment

OBJECTIVE

The objective of this experiment is to design a physicochemical treatment process that includes coagulation-flocculation, sedimentation, silica sand filtration, and activated carbon adsorption for the treatment of surface water.

INTRODUCTION

Surface water is usually treated by a physicochemical treatment process including coagulation, flocculation, filtration, and disinfection before being distributed to end users. To remove organics, color, and odors, activated carbon adsorption sometimes is also included. Typically, alum (or hydrated aluminum sulfate), ferric chloride, and ferrous sulfate are used as coagulants in this process. Recently, polymer-coagulants have also been utilized (Davis, 2010).

During the coagulation-flocculation process, $AlOH^{2+}$, $Al(OH)_2^+$, and $Al(OH)_3$ are the dominant species for aluminum salt coagulants and $Fe(OH)^+$ and $Fe(OH)_3$ are the dominant species for iron salt coagulants. Most naturally occurring colloids (e.g., microbes) are negatively charged. The positively charged species (i.e., $AlOH^{2+}$, $Al(OH)_2^+$ and $Fe(OH)^+$) are responsible for neutralizing and destabilizing the negatively charged colloids. The amorphous $Al(OH)_3$ and $Fe(OH)_3$ are responsible for clay mineral particle removal. The hydrolysis of aluminum salt coagulants or iron salt coagulants consumes alkalinity (Davis, 2010). Therefore, the coagulant addition sometimes is accompanied by a pH adjustment. Through coagulation-flocculation, flocs are formed.

During sedimentation, flocs settle to the bottom of the sedimentation tank by gravity. Once the flocs have settled in the sedimentation tank, the clear water on top will pass through filters of varying compositions (i.e., sand, anthracite, and/or garnet) and pore sizes to remove fine particles that cannot settle in the sedimentation tank. The most important parameter during filtration is head loss since it determines the backwash frequency. Head loss is a complex function of filtration rate, pressure, the influent suspended solid concentration, and the characteristics of the suspended solids and filter media. It is commonly described by the Ergun equation (Hpwe et al., 2012):

$$\frac{h_L}{L} = \frac{150\mu}{\rho g} \frac{(1-e)^2}{e^3} \frac{v_s}{(\psi d)^2} + k \frac{1-e}{e^3} \frac{v_s^2}{\psi d g}$$

where h_L is the filtration head loss, L is the length of the filter bed, μ is the absolute water viscosity, ρ is the water density, g is the gravity acceleration constant, e is the media porosity, v_s is the superficial velocity or velocity related to the surface area of the filter, ψ is the media sphericity defined as the ratio of the surface area of the equivalent volume sphere to the actual surface area of the particle, k is the Ergun coefficient for inertial loss, and d is the particle diameter. It should be noted that the Ergun equation applies to the clean bed only.

After the water is filtered, it may be treated by activated carbon adsorption to remove organic contaminants, particularly synthetic organic chemicals. Some organic chemicals cause color and odors. Activated carbon can also remove inorganic contaminants such as dadon-222, mercury, and other toxic metals (Hendricks, 2005). The adsorption process is described by Langmuir or Freundlich adsorption isotherms.

The Langmuir adsorption isotherm is mathematically expressed as (Hpwe et al., 2012):

$$q_e = \frac{Q_0 KC}{KC+1}$$

where q_e is the equilibrium contaminant concentration on the activated carbon, Q_0 and K are the Langmuir adsorption constant, and C is the equilibrium contaminant concentration in the solution.

The Freundlich adsorption isotherm is mathematically expressed as (Hpwe *et al.*, 2012):

$$\frac{x}{m} = KC^{1/n}$$

where x is the mass of adsorbed contaminant, m is the mass of adsorbent (i.e., activated carbon), C is the equilibrium concentration of adsorbed contaminant in the solution, K is the Freundlich capacity factor, and n is the Freundlich intensity parameter for a given set of adsorbate and adsorbent at a particular temperature.

Finally, a disinfectant (i.e., chlorine) is added to deactivate any remaining bacteria and to protect the water from being contaminated by germs before it is piped to homes and businesses.

MATERIALS

1. Surface water (river water or lake water)

2. Coagulant: aluminum sulfate or ferric chloride

3. Jar test apparatus

4. 1500 mL beakers

5. Silica sand

6. Activated carbon

7. Filter paper

8. Turbidity meter

9. Conical Flask

PROCEDURES

1. In this experiment, surface water will be treated by coagulation-flocculation-sedimentation, silica sand filtration, and activated carbon adsorption in series. For conducting the coagulation-flocculation-sedimentation treatment of the surface water, please refer to Experiment 5. Measure and record the turbidity and color before (influent) and after (effluent) the coagulation-flocculation-sedimentation treatment. The turbidity will be measured by the turbidity meter and color will be measured as apparent color by comparing with the standards as described in Experiment 2. Three-jar testing will be conducted to provide triple results.

2. Set up the filter using a funnel, filter paper, and silica sand. Fold the filter paper to fit the funnel following the instruction in Figure 7.1. Thoroughly wash the silica sand to remove impurities and put it on the filter paper as shown in Figure 7.2.

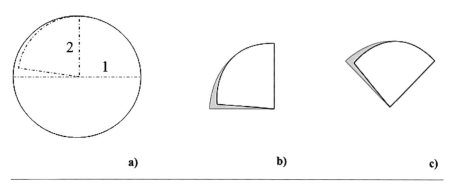

a)   b)   c)

Figure 7.1   Folding the filter paper to fit the funnel: a) → b) → C)

3. Perform the filtration of the water treated after the coagulation-flocculation-sedimentation treatment, which will serve as the influent for the filtration treatment using the setup as illustrated in Figure 7.2. Measure and record the turbidity and color before (influent) and after (effluent) the silica sand filtration treatment as described above, and record the results in Table 7.1. This will be conducted three times to provide triple results.

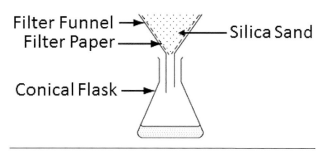

Figure 7.2   Filtration setup

4. Set up the adsorption step using a funnel, filter paper, and activated carbon. Fold the filter paper to fit the funnel following the instruction in Figure 7.1. Thoroughly rinse the activated carbon and put it on the filter paper as shown in Figure 7.3.

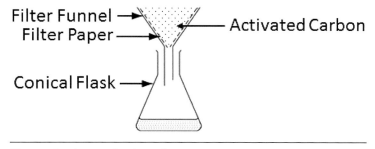

Figure 7.3   Activated carbon adsorption setup

5. Perform the activated carbon adsorption of the silica sand-filtered water, which will serve as the influent for the carbon adsorption treatment using the set up illustrated in Figure 7.3. Measure and record the turbidity and color before (influent) and after (effluent) carbon adsorption as described above in Table 7.1. Three carbon adsorption tests will be conducted to provide triple results.

DISCUSSION

1. If the river water or lake water turbidity is low, how can you improve the treatment?

2. According to your results, do you need to include activated carbon for your water treatment? Why is activated carbon adsorption needed or not needed?

3. Calculate the turbidity and color removal efficiency, E, using the following equation:

$$E = \frac{T_0 - T_1}{T_0} \times 100\%$$

4. Discuss how the media porosity, e, affects the filter head loss, h_L, by assuming reasonable parameters to plot a curve of h_L versus e.

DATA RECORDING*

Table 7.1 Turbidity and color for each treatment unit process

Treatment		Influent Turbidity (NTU)	Effluent Turbidity (NTU)	Influent Color (Color Unit)	Effluent Color (Color Unit)
Coagulation/ Flocculation and Sedimentation	Test 1				
	Test 2				
	Test 3				
Filtration	Test 1				
	Test 2				
	Test 3				
Activated Carbon Adsorption	Test 1				
	Test 2				
	Test 3				

*Triple readings are required for the measurements in this experiment. Statistical analysis is required when reporting the final results in the laboratory report.

PART 2

Hydraulic Engineering Experiments

Experiment 8

Determination of Hydraulic Head, Velocity Head, and Soil Porosity

OBJECTIVE

The objective of this experiment is for students to gain a better understanding of the concepts of Darcy's law, hydraulic head, and velocity head, as well as related calculations.

INTRODUCTION

Water flow consumes energy. Surface water flows in response to the difference of gravitational potential energy, i.e., from a high elevation to low elevation (you never see surface water naturally flowing uphill). Groundwater flow, however, is not only controlled by gravitational potential energy but also by hydraulic head energy, which includes gravitational energy in its equation. As an example, groundwater in confined aquifers may actually flow "uphill" to a higher elevation and this flow is controlled by the difference in the total energy.

The hydraulic head, or total head, is a measure of the total energy of the water, which comprises three forms, i.e., pressure energy, gravitational potential energy, and kinetic energy (White, 2015).

(i) Pressure energy:

$$W_1 = \frac{1}{m} \int_0^P V dP = \frac{1}{m} \int_0^P \frac{m}{\rho_w} dP = \frac{P}{\rho_w}$$

where W_1 is the pressure energy, m is the mass of the unit volume of water, P is the pressure, V is the unit volume of water, and ρ_w is the water density.

(ii) Gravitational energy:

$$W_2 = \frac{1}{m} \int_0^Z mg dz = gz$$

where W_2 is the gravitational energy, g is the gravity acceleration, and z is the elevation.

(iii) Kinetic energy:

$$W_3 = \frac{1}{m} \int_0^x ma dz = \frac{1}{m} \int_0^x m \frac{dv}{dt} dz = \int_0^v v dv = \frac{v^2}{2}$$

where W_3 is the kinetic energy, a is the velocity acceleration, x is the traveling distance, and v is the traveling velocity.

The total energy can be expressed as (White, 2015):

$$\Phi = \frac{P}{\rho_w} + gz + \frac{v^2}{2}$$

Eventually, to be consistent with the gauge pressure reading, the hydraulic head or total head can be expressed as:

$$H = \frac{\Phi}{g} = \frac{P}{\rho_w g} + z + \frac{v^2}{2g}$$

where $\dfrac{P}{\rho_w g}$ is the pressure head, z is the elevation head, and $\dfrac{v^2}{2g}$ is the velocity head.

Darcy's Law

In 1856, Henry Darcy examined water flow based on the results of water through beds of sand with the experimental apparatus shown in Figure 8.1. Based on his experimental observations, the fluid flow rate was proportional to the hydraulic gradient as described by Darcy's law (White, 2015):

$$Q = KA \frac{h_1 - h_2}{L}$$

where Q is the volumetric flow rate through the cylindrical column with a cross-sectional area of A and height of L, h_1 and h_2 are the hydraulic heads above the standard datum of the water in the manometer located at the input and output ports respectively, and K is the saturated hydraulic conductivity (also known as the coefficient of permeability) that describes the ease with which water can move through the sand.

Darcy's law is based on the following assumptions: (i) steady flow, (ii) constant temperature, (iii) uniform media/single fluid, (iv) incompressible fluid, (v) constant cross-sectional area, and (vi) laminar flow. The laminar flow is

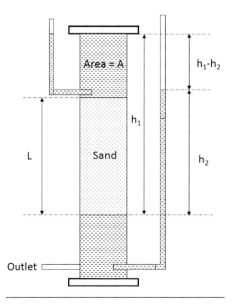

Figure 8.1   Schematic of Darcy's experiment on flow of water through sand (White, 2015)

reflected by the Reynolds number, a non-dimensional parameter that describes the balance between viscous and inertial forces (White, 2015):

$$Re = \frac{\rho v d}{\mu}$$

where Re is the Reynolds number, ρ is the fluid density, v is the fluid velocity, d is the diameter of the passageway through which the fluid travels, and μ is the dynamic viscosity of the fluid. Laminar flow is achieved when the Reynolds number is smaller than the range of 1 to 10.

Rearranging Darcy's law, we obtain Darcy's velocity: $v_D = Q/A = K(h_2-h_1)/L$. Darcy's velocity is a fictitious velocity as it assumes that flow occurs across the entire cross-section of the sediment. In fact, flow occurs through interconnected pore channels (voids), at a seepage velocity, v_S. Since the area of voids (A_v) is smaller than the area of the entire cross-section (A) (Figure 8.2), the seepage velocity ($v_S = Q/A_v$) is larger than Darcy's velocity.

Based on the continuity equation, the flow rate, Q, is constant. Subsequently, for a saturated groundwater region, the following equation is valid (Ishibashi and Hazarika, 2015):

$$Q = Av_D = A_v v_S$$

The above equation gives the relationship between v_D and v_S:

$$v_S = v_D \frac{A}{A_v}$$

Converting the area to volume by multiplying the numerator and denominator by the length, L, the equation above becomes:

$$v_S = v_D \frac{AL}{A_v L} = v_D \frac{V}{V_v} = \frac{v_D}{V_v/V} = \frac{v_D}{n}$$

where V_v is the volume of voids, V is the total volume, and n is the soil porosity (Ishibashi and Hazarika, 2015):

$$n = \frac{V_v}{V}$$

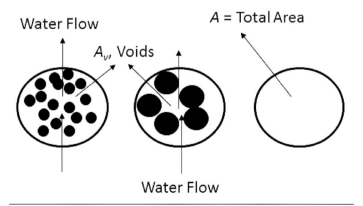

Figure 8.2   Schematic of the cross-section scenarios of the sediment

MATERIALS

1. Volumetric cylinder

2. Erlenmeyer flask

3. Ruler

4. Balance

5. Silica sand

PROCEDURES

1. Weigh an empty 25 mL volumetric cylinder (W_0).

2. Add approximately 10 mL silica sand into the volumetric cylinder, record the volume, V, and weigh the volumetric cylinder filled with silica sand, W_1, in Table 8.1. See Figure 8.3 for more details.

3. Add 10 mL water into the volumetric cylinder and record the volume, V^*.

4. Calculate the volume of the voids as follows:

$$V_v = (V + 10\ mL) - V^*$$

5. Calculate the porosity, $n = \dfrac{V_v}{V}$.

6. Repeat Steps 1-5 three times to obtain results in triplicate using fresh silica sand and record the results in Table 8.1.

7. Measure the inner diameter of the sand column, D, and calculate the cross-sectional area, A.

$$A = \frac{\pi D^2}{4}$$

8. Measure the elevation head of the sand column, h_1, as marked on the column.

9. Open the valve of the outlet and start recording the time, t, required to collected 10 mL of water using a volumetric cylinder. The flow rate, Q, can be calculated as:

$$Q = \frac{10\ mL}{t}$$

10. The seepage velocity of v_i corresponding to the elevation head of h_i is:

$$v_i = \frac{1}{n} \cdot \frac{Q_i}{A} = \frac{1}{n} \cdot \frac{10\,\dfrac{mL}{t}}{A}$$

11. Repeat Steps 7-9 for h_2 and h_3 (as marked on the column).

12. The elevation head in the sand column has the following relationship with the seepage velocity:

$$h_i = \frac{v_i^2}{2g} \quad i = 1,2,3$$

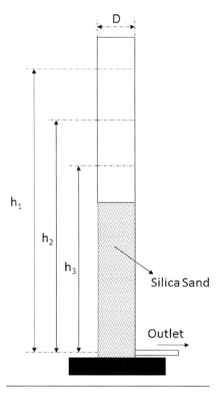

Figure 8.3   Schematic of the sand column

13. Repeat Steps 7-11 three times to obtain results in triplicate using the same silica sand column and record the results in Table 8.1.

DISCUSSION

1. According to your results of the experiment, what is the porosity of silica sand? Is the porosity related to the silica sand grain size?

2. What is the bulk density of the silica sand sample? What is the real density of the silica sand sample? Compare the two densities and check if they match the published values.

$$\rho_{bulk} = \frac{W_1 - W_0}{V}$$

$$\rho_{real} = \frac{W_1 - W_0}{V - V_v}$$

3. What is the hydraulic conductivity, K, in the sand column? Is this result reasonable?

4. If the soil of a real-world site is the same as the sand in this experiment and the hydraulic gradient is 0.01, how long will it take for the groundwater to travel 1000 m?

5. Calculate the Reynolds number using the seepage velocity obtained. Discuss Darcy's law assumptions and validity.

6. Based on the elevation head, is the calculated seepage velocity consistent with the measured seepage velocity in the experiment?

DATA RECORDING*

Table 8.1 Sand column experimental results

Parameter		Readings
W_0		
W_1		
V		
V^*		
D		
h_0	t	
h_1	t	
h_2	t	

*Triple readings are required for the measurements in this experiment. Statistical analysis is required when reporting the final results in the laboratory report.

Experiment 9

Measurement of Air Pressure in a Closed System

OBJECTIVE

The objective of this experiment is to measure the air pressure in a closed system (system pressure).

INTRODUCTION

As discussed before, the hydraulic head of a water column is composed of the pressure head, $\dfrac{P}{\rho_w g}$, elevation head, z, and the velocity head, $\dfrac{v^2}{2g}$ (White, 2015):

$$H = \frac{P}{\rho_w g} + h + \frac{v^2}{2g}$$

where h is the elevation of the water column above the datum, P is the pressure of the system, ρ_w is the density of water, g is the gravity acceleration, and v is the velocity.

Under static conditions, the velocity head is zero. Thus, the hydraulic head is the sum of the pressure head and the elevation head:

$$H = \frac{P}{\rho_w g} + h$$

For a closed system, as shown in Figure 9.1, the system pressure, P, is balanced by the water head ($\rho_w g h$) and the atmospheric pressure, P_0:

$$P = P_0 + \rho_w g h$$

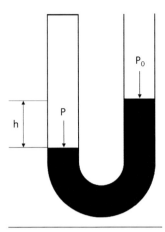

Figure 9.1 Demonstration of the total pressure as the sum of the water head and atmospheric pressure

MATERIALS

1. Pressure meters

2. Ruler

3. The hydrostatic pressure apparatus is shown in Figure 9.2. In the hermetically sealed water column, five pressure meters are arranged along the depth of the column, providing readings of the hydraulic head (sum of pressure head and elevation head) at the corresponding depth. The system pressure can be adjusted by pumping air into the system, which changes the readings of the pressures meters.

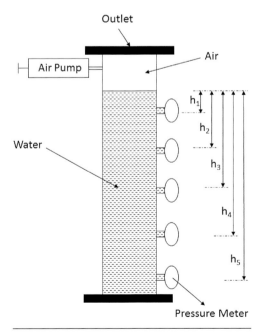

Figure 9.2 Schematic of the hydrostatic pressure apparatus for the experiment

PROCEDURES

1. Before the experiment, the column is filled with trapped air in the top. Open the outlet of the column to make the system pressure equal to the atmospheric pressure.

2. Zero all the pressure meters and reconnect them to the column.

3. Read the pressure meters at the different depths and record the readings in Table 9.1.

4. Pump air in to increase the system pressure (i.e., Level 1).

5. Read the pressure meters and record the readings in Table 9.1.

6. Keep increasing the system pressure to the next two levels (i.e., Level 2 and Level 3) by pumping in more air, read the pressure meters, and record the readings in Table 9.1.

7. Measure the elevations at h_1, h_2, h_3, h_4, and h_5.

8. Calculate the air pressures in the four scenarios (no air introduced and three levels of air pressure) and record the results in Table 9.2. Discuss the uniformity of the results estimated from the five pressure meters.

$$P = Preasure\ Meter\ Reading - \rho_w gh$$

9. Repeat Steps 1-8 to get measurements/readings in triplicate and record the results in Tables 9.1 and 9.2.

DISCUSSION

1. What is the calculated system pressure before the introduction of air? Is it the same as the atmospheric pressure (approximately 14.696 psi)?

2. Compare the results of the system pressure calculated from pressure meter readings at different locations at each system pressure level. Report the standard deviations for your discussion.

3. Discuss any possible interference concerning the pressure meter readings.

DATA RECORDING*

Table 9.1 Pressure meter readings when different air pressures are added into the air chamber

Pressure Meter #	No Air Introduced	System Pressure Level 1	System Pressure Level 2	System Pressure Level 3
1				
2				
3				
4				
5				

Table 9.2 Calculated system pressures based on different pressure meter readings

Pressure Scenario	Pressure Meter 1	Pressure Meter 2	Pressure Meter 3	Pressure Meter 4	Pressure Meter 5
No Air Introduced					
System Pressure Level 1					
System Pressure Level 2					
System Pressure Level 3					

*Triple readings are required for the measurements in this experiment. Statistical analysis is required when reporting the final results in the laboratory report.

Experiment 10

Buoyant Force Measurement

OBJECTIVE

In this experiment, buoyant force will be measured for a curved block using the laboratory apparatus. The results will be compared with that of theoretical calculations.

INTRODUCTION

Buoyant force, exerting on an object wholly or partly immersed in water, is calculated based on the Archimedes' principle. It equals to the weight of water displayed by the object (White, 2015):

$$F_b = W_w = \rho_w g V$$

where F_b is the buoyant force, W_w is the weight of the displayed water, ρ_w is the density of water, g is the gravity acceleration, and V is the displayed water volume.

The buoyant force that acts on the immersed object can be calculated at rotational equilibrium when the sum of the torques is equal to zero using the laboratory apparatus shown in Figure 10.1. This apparatus contains a curved block in a Plexiglass box, the weight of which can be balanced by the buoyant force and added weight based on the torque balance.

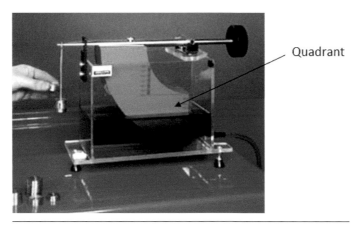

Figure 10.1 Photograph of laboratory apparatus for buoyant force measurement

The pressures on the curved surfaces do not cause a moment about the support point since they are a normal stress. In addition, the pressures on the two opposite plane sides of the quadrant cancel each other and the buoyant force acts upwards on the plane surface of the quadrant. The torque force of the added weight is balanced with the buoyant force at the point of support on the knife edge.

MATERIALS

1. The buoyancy measurement apparatus

2. Water

3. Balancing weight kits

PROCEDURES

1. Check the depth label of the Plexiglass box. Make sure it is clearly marked for each depth increment. The depth and other dimensions of the box will be used to calculate volumes in the following steps.

2. Balance the quadrant by adjusting the counterbalance as shown in Figure 10.1.

3. Add known amount of water, V_1, to the Plexiglass box. Record the depth of water in the Plexiglass, calculate the volume, V_2, and record the results in Table 10.1.

4. Balance the quadrant by adding weight to the weight hanger. Record the added weight, W, in Table 10.1

5. Subtract V_2 from V_1 to get the immersed volume of the quadrant, V_3.

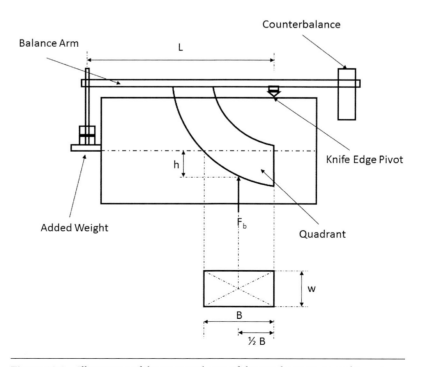

Figure 10.2   Illustration of the projected area of the quadrant immersed in water

6. Calculate the buoyant force by:

$$F_b = \rho_w g V_3$$

7. Balancing the moments about the pivot gives (Figure 10.2):

$$WL = F_b \cdot \frac{1}{2} B$$

where L is the horizontal distance from the knife edge pivot to the centroid of the added weight.

8. Empty the Plexiglass box. Repeat Steps 3-7 two more times by adding a different known volume of water to the Plexiglass box to obtain observations in triplicate and record the results in Table 10.1.

DISCUSSION

1. How does B vary with respect to the volume of added water?

2. In Step 3 of the procedure, the quadrant may be partially submerged or fully submerged, depending on how much water you add. How would this affect your experiment?

DATA RECORDING*

$L =$ _____

$W =$ _____

Table 10.1 Buoyant force experimental results

V_1	V_2	W	V_3	$F_b = \rho_w g V_3$

*Triple readings are required for the measurements in this experiment. Statistical analysis is required when reporting the final results in the laboratory report.

Experiment 11

Osborne Reynolds Fluid Flow

OBJECTIVE

In this experiment, the three types of flow (i.e., laminar, trasient, and turbulent) will be simulated using a dye for observation. The type of flow is determined by the corresponding Reynolds number.

INTRODUCTION

In fluid mechanics, the Reynolds number, Re, is a dimensionless ratio of inertial forces to viscous forces for a given flow condition. It is commonly used to predict flow patterns (White, 2015). The Re can be expressed as:

$$Re = \frac{\rho u L}{\mu} = \frac{uL}{v}$$

where ρ is the liquid density, u is the flow velocity, L is the characteristic length, μ is the fluid dynamic viscosity, and v is the fluid kinematic viscosity. Re is correlated with the velocity of the flow, u. Given the same fluid and pipe or duct, the faster the fluid flows, the larger the Re is.

For a pipe or duct, the characteristic length is the hydraulic diameter, d_h, which can be calculated by:

$$d_h = \frac{4A}{P} = \frac{4\pi r^2}{2\pi r} = 2r = d$$

where A is the cross-section area of the pipe or duct, P is the wetted perimeter, r is the radius of the pipe or the duct, and d is the diameter of the pipe or duct.

The following Re range can be used to determine the flow type:

- $Re < 2300$ Laminar flow
- $2300 < Re < 4000$ Transient flow
- $Re > 4000$ Turbulent flow

Laminar flow generally ocuurs in small pipes and at low flow velocities. Turbulent flow occurs in large pipes and at high flow velocities. Transitional flow is a flow type between laminar and turbulent flows. Each of these flow types behaves in different manners in terms of frictional energy loss. A schematic of streamlines in the three types of flow is shown in Figure 11.1. Laminar flow has stable and straight streamlines; turbulent flow has unstable and irregular streamlines; and transient flow is a mixture of the two.

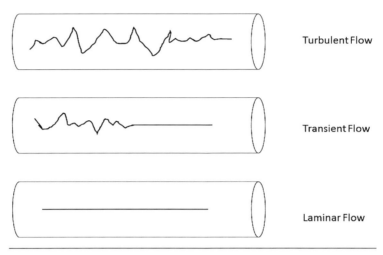

Turbulent Flow

Transient Flow

Laminar Flow

Figure 11.1   Schematic of streamlines in the three flow types

MATERIALS

1. Osborne Reynolds flow demonstration apparatus (Figure 11.2)

    This apparatus is used to display laminar and turbulent flows. In the experiment, it is possible to observe the transition from laminar to turbulent flow by varying the flow rate. The *Re* number is used to assess whether a flow is laminar or turbulent.

Figure 11.2   Osborne Reynolds flow demonstration apparatus

This apparatus consists of a Acrylic tube with a flow-optimized inlet, through which water flows. A valve is attached to adjust the flow rate. Ink is injected into the flowing water through a injection port. A layer of glass beads in the water tank helps to produce an even and low-turbulent flow.

2. Ink

PROCEDURES

1. Start with one slow flow rate and wait for the flow to become stable. This flow should be a laminar flow with $Re < 2300$.

2. Release the ink from the separatory funnel into the flowing pipe.

3. Observe the flow of the streamline in the pipe and record the diameter, D, of the flow regime using a ruler (see Figure 11.3). Record your data in Table 11.1.

4. Stop releasing the ink until no color is observed.

5. At this flow rate range, repeat the above steps two more times to obtain the observations in triplicate and record the results in Table 11.1.

6. Increase the flow rate to make a transient flow with $2300 < Re < 4000$. Repeat Steps 2-5 three times to obtain the observations in triplicate and record the results in Table 11.1.

7. Keep increasing the flow rate to make a turbulent flow with $Re > 4000$. Repeat Steps 2-5 three times to obtain the observations in triplicate and record the results in Table 11.1.

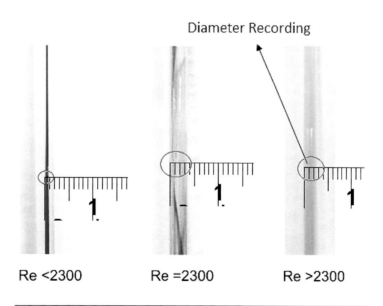

Figure 11.3   Observational flow streamlines in the apparatus

DISCUSSION

1. Compare your observations with Figure 11.3 to determine the *Re* number in your tests and the corresponding flow flow types.

2. Based on the *D* values measured in Step 3, calculate the hydraulic diameter, d_h, and the *Re* numbers, and then estimate the ranges of the flow velocities, *u*, according to the flow types that you observe.

DATA RECORDING*

Table 11.1   Flow velocity, rate, type, and *Re* number

Observation	Flow Rate	Flow Type	*Re* Number	Velocity
1				
2				
3				
4				
5				
6				
7				
8				

*Triple readings are required for all measurements in this experiment. Statistical analysis is required when reporting the final results in the laboratory report.

Experiment 12

Bernoulli's Theorem

OBJECTIVE

In this experiment, Bernoulli's Theorem will be tested in a Venturi tube.

INTRODUCTION

Bernoulli's Equation

At any arbitrary point along a streamline for an incompressible flow, the sum of velocity head, elevation head, and pressure head is a constant. This relationship is described by Bernoulli's equation (White, 2015):

$$\frac{v^2}{2} + gz + \frac{p}{\rho} = constant$$

where v is the velocity of the fluid at a point on a streamline, g is the gravity acceleration, z is the elevation of the point above a reference plane, p is the pressure at the chosen point, and ρ is the density of the fluid.

It should be noted that Bernoulli's equation is based on the following assumptions:

- Flow must be incompressible—even though pressure varies, the density must remain constant along a streamline;

- Flow must be steady, i.e., the fluid velocity at a point cannot change with time; and

- Friction by viscous forces has to be negligible.

Venturi Tube

A Venturi tube is a device used to demonstrate Bernoulli's Theorem. It consists of a tube with a short, narrow center section and widened, tapered ends (Figure 12.1). As the fluid moves through, it accelerates in the direction

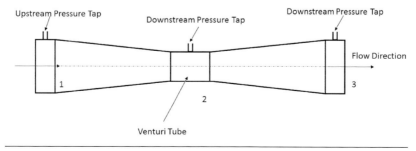

Figure 12.1   Schematic of a Venturi Tube

of the tapering contraction with an increased velocity in the throat. Owing to the increased velocity, the pressure drops, following Bernoulli's equation. The pressure along the streamline can be measured by piezometers, which can reflect the changes of velocity at the throat. This pressure change is termed the Venturi effect (White, 2015).

The pressure head at point 1 is higher than that of point 2 and the fluid velocity at point 1 is lower than that of point 2 because the cross-sectional area at point 1 is greater than of point 2. Based on Bernoulli's equation, the pressure drop at the center streamline can be related to the change of velocity as follows:

$$\Delta P = p_1 - p_2 = \frac{\rho}{2}(v_2^2 - v_1^2)$$

where ρ is the density of the fluid, v_1 is the fluid velocity at point 1, and v_2 is the fluid velocity at point 2.

A Venturi tube can also be used to measure the volumetric flow rate, Q, according to the continuity equation:

$$Q = v_1 A_1 = v_2 A_2$$

where A_1 and A_2 are the areas of the cross section at points 1 and 2, respectively. Q thus can be calculated as:

$$Q = A_1 \sqrt{\frac{2}{\rho} \cdot \frac{\Delta P}{\left(\frac{A_1}{A_2}\right)^2 - 1}} = A_2 \sqrt{\frac{2}{\rho} \cdot \frac{\Delta P}{1 - \left(\frac{A_2}{A_1}\right)^2}}$$

Based on Bernoulli's equation, at any two points, 1 and 2,

$$\frac{v_1^2}{2} + gz + \frac{p_1}{\rho} = \frac{v_2^2}{2} + gz + \frac{p_2}{\rho}$$

Along the streamline, the elevation head stays the same. If the area of cross section at point 2 is much smaller than that of point 1, velocity at point 1 is negligible. Hence, the above equation can be simplified as:

$$\frac{v_2^2}{2} + \frac{p_2}{\rho} = \frac{p_1}{\rho}$$

The velocity at point 2 can thus be calculated as:

$$v_2 = \sqrt{\frac{2\Delta P}{\rho}}$$

MATERIALS

1. Venturi tube

2. Monometer

PROCEDURES

1. Measure and record the diameter of the pipe, D, along the streamline where the three pressure taps are located and then calculate and record the cross-sectional areas, A, in the blanks above Table 12.1 according to the following equation:

$$A = \frac{\pi D^2}{4}$$

2. Fix the flow rate of the Venturi tube to Q_1 and record monometer reading at each pressure tap in Table 12.1. Calculate the pressure difference, ΔP and record the results in Table 12.2.

3. Calculate Q, v_1 and v_2 and record the results in Table 12.3.

4. Change the flow rate of the Venturi tube to Q_2, Q_3 and Q_4, repeat the above measurements and calculations and record the results in Table 12.1, Table 12.2 and Table 12.3 respectively.

5. Repeat Steps 2 to 4 two more times to obtain triplicate results for flow rates of Q_1, Q_2, Q_3 and Q_4 and record the results in Table 12.1, Table 12.2 and Table 12.3 respectively.

DISCUSSION

1. Is the recorded flow rate consistent with the calculated one? Comment on the discrepancy and interference, if any.

2. Check the continuity equation using the recorded and calculated flow rates. Comment on the discrepancy and interference, if any.

3. What assumptions are made for the calculation of Q and v?

DATA RECORDING*

Section 1   Diameter, D _____,   Area, A

Section 2   Diameter, D _____,   Area, A

Section 3   Diameter, D _____,   Area, A

Table 12.1 Monometer readings when different flow rates are applied (unit: mm)

Monometer #	Flow Rate 1	Flow Rate 2	Flow Rate 3	Flow Rate 4
1 (Section 1)				
2 (Section 2)				
3 (Section 3)				

Table 12.2 Pressure differences, ΔP

Monometer #	Flow Rate 1	Flow Rate 2	Flow Rate 3	Flow Rate 4
1 & 2				
2 & 3				

Table 12.3 Calculated Q and v

	Flow Rate 1	Flow Rate 2	Flow Rate 3	Flow Rate 4
Q				
v_1				
v_2				
v_3				

*Triple readings are required for the measurements in this experiment. Statistical analysis is required when reporting the final results in the laboratory report.

Experiment 13

Head Loss in Pipes

OBJECTIVE

In this experiment, head loss will be investigated when water flows through different fittings and in long pipes with variable diameters.

INTRODUCTION

Friction Head Loss

When water flows in a long pipe, a portion of energy is lost due to friction resistance along the pipe and the fittings, which is termed as friction head loss. Figure 13.1 illustrates the head loss due to friction in a pipe (White, 2015).

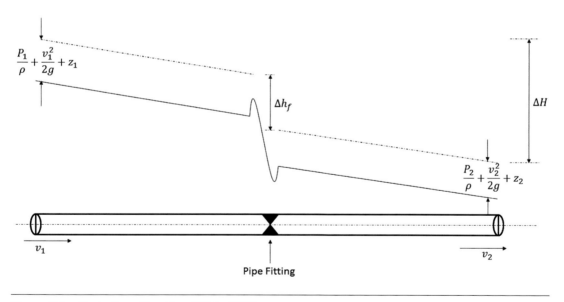

Figure 13.1   Illustration of head loss (White, 2015)

The total head loss, ΔH, is the sum of the head loss due to passing through the pipe fittings, h_f, and head loss due to skin friction, h_L (White, 2015):

$$\Delta H = h_f + h_L = \left(\frac{P_1}{\rho} + \frac{v_1^2}{2g} + z_1 \right) - \left(\frac{P_2}{\rho} + \frac{v_2^2}{2g} + z_2 \right)$$

where v_u and v_d are the upstream and the downstream velocity, respectively.

If there is no change in elevation and pipe diameter along the streamline, the total head loss can be calculated as:

$$\Delta H = h_f + h_L = \frac{P_1}{\rho} - \frac{P_2}{\rho} = \frac{\Delta P}{\rho}$$

The Darcy-Weisbach equation is commonly used to describe the head loss of pipe friction, h_L (White, 2015):

$$h_L = f \frac{L}{D} \frac{v^2}{2g}$$

where f is the friction factor, L is the length of the pipe, D is the diameter of the pipe, v is the flow velocity, and g is the gravity acceleration. The dimensionless friction factor, f, is a function of the pipe diameter, D, the roughness of the pipe surface, ε, and the Reynolds number, Re (White, 2015):

$$f = f\left(Re, \frac{\varepsilon}{D}\right)$$

In practice, the Moody chart is commonly used to calculate f. For laminar flow (i.e., $Re < 2300$), $f = \dfrac{64}{Re}$. For turbulent flow (i.e., $Re > 4000$), f is not a function of Re, but a function of ε and D, such that $f = f\left(\dfrac{\varepsilon}{D}\right)$. f in transition and turbulent flows can be calculated by the Colebrook equation (White, 2015):

$$\frac{1}{\sqrt{f}} = -2.0\log\left(\frac{\varepsilon/D}{3.7} + \frac{2.51}{Re\sqrt{f}}\right)$$

The Colebrook equation is implicit in f and its determination requires tedious iteration. An explicit relation is given by Haaland as (White, 2015):

$$\frac{1}{\sqrt{f}} = -1.8\log\left[\frac{6.9}{Re} + \left(\frac{\varepsilon/D}{3.7}\right)^{1.11}\right]$$

Head loss of water flowing through the pipe fittings, h_f, including pipe entrance or exit; sudden expansion or contraction; bends, elbows, tees, and other fittings; valves, open or partially closed; and gradual expansions or contractions, can be calculated by:

$$h_L = K_L \frac{v^2}{2g}$$

where K_L is loss coefficient for pipe fittings.

MATERIALS

1. Fluid friction apparatus with attached monometers (Figure 13.2)

   The fluid friction apparatus has four parallel pipe lines that are different in diameter (top four pipe lines in Figure 13.2) and a pipe line with variable fittings (the bottom pipe line at the bottom of Figure 13.2). The flow can be controlled to pass through one of these pipe lines and the flow rate can be adjusted by the main valve. The head loss through the pipes and fittings can be measured when a steady flow is reached. To measure the head loss, hook the two ends of the water manometer and mercury manometer to the two piezometers across

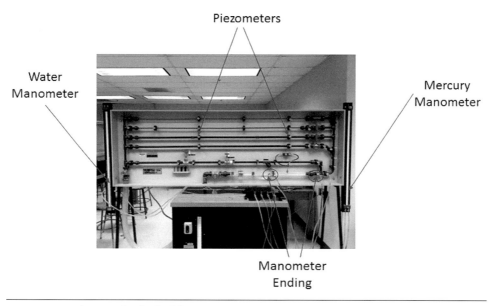

Figure 13.2 Photograph of a fluid friction apparatus

the section of interest (Figure 13.2). The levels in the two limbs of the manometer give the pressure difference between these two piezometers. When water manometer is used, the head loss between these two piezometers can be calculated by:

$$\Delta H = h_f + h_L = \frac{P_1}{\rho_w} - \frac{P_2}{\rho_w} = \frac{\Delta P}{\rho_w}$$

where ρ_w is the water density.

Similarly, when the mercury manometer is used, the head loss between the two piezometers is:

$$\Delta H = h_f + h_L = \frac{P_1}{\rho_{Hg}} - \frac{P_2}{\rho_{Hg}} = \frac{\Delta P}{\rho_{Hg}}$$

where ρ_{Hg} is the mercury density.

PROCEDURES

1. Measure and record the diameter and length of the four pipes in the blanks above Table 13.1.

2. Start with a small flow rate that corresponds to a laminar flow. Record the flow rate in Tables 13.1 and 13.2 by reading the flow meter. Hook the two ends of the water manometer and mercury monometer to the two piezometers of the four pipes to record the head loss across the two piezometers. Record your data in Tables 13.1 and 13.2. Note that a 1 mm differential reading in the mercury manometer represents a 12.6 mm differential reading in the water manometer. Calculate the Reynolds number, Re, which should be smaller than 2000. It is recommended that the manometers are read three times and a mean value of a differential head (head loss) is recorded.

3. Similarly, hook the two ends of the water manometer and mercury monometer to the two piezometers of one valve and one of the other fittings (such as bends, elbows, or tees) of your choosing to record the head loss of the flow passing. Record your data in Tables 13.1 and 13.2.

4. Increase the flow rate seven times and repeat the measurement and calculation in Steps 2 and 3 after each flow rate increase. Record the results in Tables 13.1 and 13.2. The eight flow rates should cover three flow types: laminar, transition, and turbulent flow, with at least two flow rates in each flow type. Plot a graph of head loss and *Re* number against flow rate as the experiment proceeds to help you determine the flow rate increase.

5. Calculate *f* as a function of velocity, *v*, and plot *f* versus *Re*. Quantify the head loss for water flow in pipes and when passing through fittings at different flow velocity. Record related data in Table 13.2.

DISCUSSION

1. Calculate the head losses for the known flow rates. What is the relationship between the head loss and the flow rate?

2. Perform the measurements for the two pipes with the same dimensions but different roughness, and plot the friction factor against the *Re* number. Is the trend of your plot consistent with that of the Moody diagram? Comment on the discrepancy, if any.

DATA RECORDING

Pipe 1   D = _____,   L = _____

Pipe 2   D = _____,   L = _____

Pipe 3   D = _____,   L = _____

Pipe 4   D = _____,   L = _____

D: Diameter   L: Length

Table 13.1   Recording of the monometer reading

#	Flow Rate	Water Monometer Reading Difference (mm)					
		Pipe 1	Pipe 2	Pipe 3	Pipe 4	Valve	Other Fitting
1							
2							
3							
4							
5							
6							
7							
8							

#	Flow Rate	Mercury Monometer Reading Difference (mm)					
		Pipe 1	Pipe 2	Pipe 3	Pipe 4	Valve	Other Fitting
1							
2							
3							
4							
5							
6							
7							
8							

Table 13.2   Data processing

Pipe 1

Flow Rate	1	2	3	4	5	6	7	8
Q								
v								
Δh								
ΔP								
h_f								
f								
Re								

Pipe 2

Flow Rate	1	2	3	4	5	6	7	8
Q								
v								
Δh								
ΔP								
h_f								
f								
Re								

Pipe 3

Flow Rate	1	2	3	4	5	6	7	8
Q								
v								
Δh								
ΔP								
h_f								
f								
Re								

Pipe 4

Flow Rate	1	2	3	4	5	6	7	8
Q								
v								
Δh								
ΔP								
h_f								
f								
Re								

Valve

Flow Rate	1	2	3	4	5	6	7	8
Q								
v								
Δh								
ΔP								
h_f								
f								
Re								

Other Fitting

Flow Rate	1	2	3	4	5	6	7	8
Q								
v								
Δh								
ΔP								
h_f								
f								
Re								

Experiment 14

Hydraulic Flume (Open Channel Flow)

OBJECTIVE

This experiment will evaluate energy loss and generate the specific force and energy curves by hydraulic jumps in a hydraulic flume.

INTRODUCTION

A hydraulic jump occurs when a liquid with a high velocity is discharged into a zone with a lower velocity (Figure 14.1). This is a common phenomenon in an open channel. The reduction of velocity leads to an increase in liquid elevation since the kinetic energy of the liquid converts to potential energy. Note that some of the energy is dissipated in the form of heat due to turbulence.

Hydraulic jump occurs when there is a flow impediment downstream, such as a weir, bridge abutment, dam, or simply channel friction (White, 2015). During hydraulic jump, water depth increases and energy is dissipated as turbulence. When hydraulic jump occurs, flow will go from supercritical (i.e., water depth smaller than the critical depth where energy is at a minimum) to subcritical (i.e., water depth greater than the critical depth),

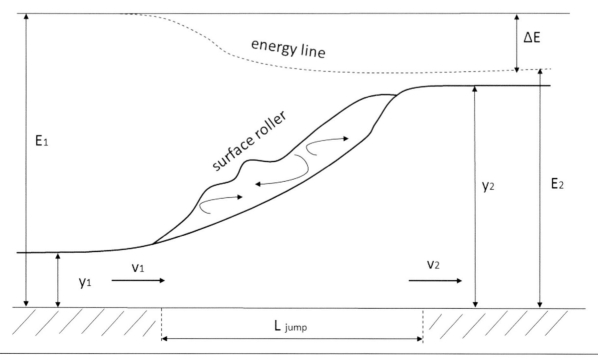

Figure 14.1 Example diagram of hydraulic jump (White, 2015)

93

and the Froude Number, $Fr = \dfrac{v}{\sqrt{gL}}$, will change from greater than 1 (i.e., $Fr = \dfrac{v_1}{\sqrt{gL}} > 1$) to smaller than 1 (i.e., $Fr = \dfrac{v_2}{\sqrt{gL}} < 1$). Fr is a dimensionless number defined as the ratio of the flow inertia to the external field

(due to gravity in this case). The hydraulic jump can be classified based on initial Fr as undular, weak, oscillating jet, steady, and strong (Table 14.1).

Table 14.1   Hydraulic types

	Froude Number	Observations		
Undular Jump	1.0 to 1.7	Low energy dissipating rate		Smooth downstream water surface
Weak Jump	1.7 to 2.5			
Oscillating Jet Jump	2.5 to 4.5	Irregular fluctuation		Turbulence downstream
Steady Jump	4.5 to 9.0	Jumps at some locations		Turbulence within the jump
Strong Jump	> 9.0	High energy dissipation rate		Large change in depth

The uniform flow rate in an open channel can be calculated by the empirical Manning's equation (White, 2015):

$$Q = CR^{1/2}S_e^{1/2}A = \frac{1}{n}R^{2/3}S_e^{1/2}A$$

where Q is the flow rate, $C = \dfrac{1}{n}R^{1/6}$ is the Chezy coefficient, R is the hydraulic radius, S_e is the slope of the total energy line, A is the cross-section area, and n is the Manning's roughness coefficient. R can be calculated by:

$$R = \frac{A}{P} = \frac{b \cdot y_{avg}}{b + 2y_{avg}}$$

where P is the wetted perimeter, b is the width of the channel, and y_{avg} is the average water depth of the upstream water depth, y_1, and downstream water depth, y_2. S_e can be quantified by the following equation:

$$S_e = \frac{(y_1 + \dfrac{v_1^2}{2g}) - \left(y_2 + \dfrac{v_2^2}{2g}\right)}{L} + S_0$$

where L is the length between the upstream and downstream, v_1 is the upstream flow velocity, v_2 is the downstream flow velocity, and S_0 is the bottom slope. For uniform flow, S_0 is the same as the slope of the energy grade line and the water surface slope.

The upstream energy is:

$$E_1 = y_1 + \frac{v_1^2}{2g}$$

The head loss during water jump is:

$$h_L = \frac{(y_2 - y_1)^8}{4 y_1 y_2}$$

MATERIALS

1. Laboratory hydraulic flume

2. Flow rate meter

PROCEDURES

1. Turn on the pump and run the hydraulic flume until a steady flow is reached. Note that a hump has been arranged in the flume by the instructor.

2. Use a ruler to measure the upstream and downstream water levels, y, the length between the upstream and downstream, L, and the width, b. Record the measurements in Table 14.2.

3. Use the flow meter to measure the upstream and downstream flow rate, v, and record the results in Table 14.2.

4. Change the flow rate, Q, and repeat the above measurements seven more times and record the results in Table 14.2.

DISCUSSION

1. Calculate the Froude number and determine the type of hydraulic jumps you observe using the Table 14.1.

2. Calculate the Chezy coefficient for all the runs.

3. Plot the variation in Chezy coefficient versus the water depth (tail gate position) for each flow rate.

DATA RECORDING

Table 14.2 Open channel flow experimental results

Q	Section	b	L	y	v	S_e	y_{ave}	A	P	R	C	n
	1 (upstream)											
	2 (downstream)											
	1											
	2											
	1											
	2											
	1											
	2											
	1											
	2											
	1											
	2											
	1											
	2											
	1											
	2											

References

Andreadis, A., 2002. *An Optical Nephelometric Model Design Method for Particle Characterization*. Loughborough University, Loughborough.

Berger, P.S., Argaman, Y., 1983. *Assessment of Microbiology and Turbidity Standards for Drinking Water*. EPA, Washington D.C.

Boyd, C.E., Tucker, C.S., 1998. *Pond Aquaculture Water Quality Management*. Kluwer Academic, Boston, MA.

Davis, M., 2010. *Water and Wastewater Engineering*. McGraw-Hill, New York, NY.

Droste, R.L., 1996. *Theory and Practice of Water and Wastewater Treatment*. John Wiley & Sons, Inc., Danvers, MA.

Eaton, A.D., Clesceri, L.S., Rice, E.W., Greenberg, A.E., 2005. *Standard Methods for the Examination of Water and Wastewater*. American Public Health Association, Washington D.C.

Fielding, M., Farrimond, M., 1999. *Disinfection By-Products in Drinking Water: Current Issues*. Royal Society of Chemistry, Cambridge CB40WF, UK.

Gilvear, D.J.A., 1987. *Suspended Solids Transport Dynamics in Regulated Rivers*. Loughborough University.

Hach, 2016. *Hach Methods Quick Reference Guide*. Hach World Headquarters, Loveland, CO.

Hendricks, D.W., 2005. *Water Treatment Unit Processes: Physical and Chemical*. Taylor & Francis, Boca Raton, FL.

Henze, M., Harremoes, P., la Cour Jansen, J., Arvin, E., 2002. *Wastewater Treatment: Biological and Chemical Processes*. Springer, New York, NY.

Hpwe, K.J., Hand, D.W., Crittenden, J.C., Trussell, R.R., Tchobanoglous, G., 2012. *Principles of Water Treatment*. John Wiley & Sons, New York, NY.

Ishibashi, I., Hazarika, H., 2015. *Soil Mechanics Fundamentals and Applications*. CRC Press, Boca Raton, FL.

Jackson, M.L., 1969. *Soil Chemical Analysis: Advanced Course*. UW-Madison Libraries Parallel Press, Madsion, WI.

McGuire, M.J., 2013. *Chlorine Revolution: The History of Water Disinfection and the Fight to Save Lives*. American Water Works Association, Denver, CO.

Morel, F.M.M., Hering, J.G., 1993. *Principles and Applications of Aquatic Chemistry*. John Wiley & Sons, Inc., New York, NY.

Russell, S., 1994. *Turbidity: A Guide to Measurement in Water Applications*. WRS, Waco, TX.

USEPA, 1977. *National Primary Drinking Water Regulations*. CFR Title 40. Washington D.C.

USEPA, 2009. *National Recommended Water Quality Criteria*. Washington D.C.

White, F., 2015. *Fluid Mechanics*. McGraw-Hill, New York, NY.